T0224324

# Digital Heritage Reconstruction Using Super-resolution and Inpainting

# Synthesis Lectures on Visual Computing
## Computer Graphics, Animation, Computational
## Photography, and Imaging

Editor
**Brian A. Barsky**, *University of California, Berkeley*

This series presents lectures on research and development in visual computing for an audience of professional developers, researchers, and advanced students. Topics of interest include computational photography, animation, visualization, special effects, game design, image techniques, computational geometry, modeling, rendering, and others of interest to the visual computing system developer or researcher.

Digital Heritage Reconstruction Using Super-resolution and Inpainting

Milind G. Padalkar, Manjunath V. Joshi, and Nilay L. Khatri

ISBN: 978-3-031-01463-5      paperback
ISBN: 978-3-031-02591-4      ebook

DOI 10.1007/978-3-031-02591-4

A Publication in the Springer series
*SYNTHESIS LECTURES ON VISUAL COMPUTING: COMPUTER GRAPHICS, ANIMATION, COMPUTATIONAL PHOTOGRAPHY, AND IMAGING*

Lecture #26
Series Editor: Brian A. Barsky, *University of California, Berkeley*
Series ISSN
Print 2469-4215    Electronic 2469-4223

# Digital Heritage Reconstruction Using Super-resolution and Inpainting

Milind G. Padalkar
Dhirubhai Ambani Institute of Information and Communication Technology (DA-IICT), India

Manjunath V. Joshi
Dhirubhai Ambani Institute of Information and Communication Technology (DA-IICT), India

Nilay L. Khatri
Jekson-Vision, India

*SYNTHESIS LECTURES ON VISUAL COMPUTING: COMPUTER GRAPHICS, ANIMATION, COMPUTATIONAL PHOTOGRAPHY, AND IMAGING #26*

# ABSTRACT

Heritage sites across the world have witnessed a number of natural calamities, sabotage and damage from visitors, resulting in their present ruined condition. Many sites are now restricted to reduce the risk of further damage. Yet these masterpieces are significant cultural icons and critical markers of past civilizations that future generations need to see. A digitally reconstructed heritage site could diminish further harm by using immersive navigation or walkthrough systems for virtual environments. An exciting key element for the viewer is observing fine details of the historic work and viewing monuments in their undamaged form. This book presents image super-resolution methods and techniques for automatically detecting and inpainting damaged regions in heritage monuments, in order to provide an enhanced visual experience.

The book presents techniques to obtain higher resolution photographs of the digitally reconstructed monuments, and the resulting images can serve as input to immersive walkthrough systems. It begins with the discussion of two novel techniques for image super-resolution and an approach for inpainting a user-supplied region in the given image, followed by a technique to simultaneously perform super-resolution and inpainting of given missing regions. It then introduces a method for automatically detecting and repairing the damage to dominant facial regions in statues, followed by a few approaches for automatic crack repair in images of heritage scenes. This book is a giant step toward ensuring that the iconic sites of our past are always available, and will never be truly lost.

# KEYWORDS

super-resolution, inpainting, cultural heritage, crack detection, digital reconstruction

*To*

*My mother Suniti and my father Gajanan*
*— MGP*

*Smita, Nidhi and Ninad*
*— MVJ*

*My mother Niharika, My father Lalit,*
*My brother Prerak, and Bajaru*
*— NLK*

# Contents

# Preface

Over the past few years, we have witnessed an active participation of many global organizations toward the preservation of cultural heritage. Research in this area is being disseminated through a series of articles in journals, conferences and workshops that focus on the preservation and digital reconstruction of historic monuments. The tools and techniques developed by the vision and graphics community are finding wider applications in this area. Moreover, new techniques are being tailored for the applications involving cultural heritage. This is a sign that initiatives in this area are receiving a greater acceptance among researchers.

With the availability of better facilities for computing, storage and communication systems, we notice that there is a gradual transition toward the development of immersive walkthrough systems and virtual museums. In immersive walkthrough systems, tourists can experience being at a heritage site without physically visiting the site. They can feel an exciting onsite experience with mixed reality by using their handheld devices, with which they can peek into the historic view of the heritage sites. The tourists may be provided a digitally reconstructed view, allowing them to view damaged monuments in their entirety, complete with the skillful work that may have existed prior to any destruction. This possibility drives us toward research in the areas of super-resolution and automatic inpainting for digital heritage reconstruction.

The existing literature and reference books on the topic of digitalization for cultural heritage focus on providing information about the required hardware and software setup, acquisition of images for 2D and 3D rendering, preservation of original objects at the heritage site or in museums as well as creating their imaged replica in digital form, etc. However, the other aspect of digitizing the cultural heritage viz. reconstruction and recovery of details in the lost or deteriorated regions in the photographs of the monuments has received less attention, and the amount of published literature under this category is very much limited.

In this book we discuss this aspect of digitizing the cultural heritage and present methods for image super-resolution and techniques for automatically detecting and inpainting regions like cracks and other damaged regions in heritage monuments. The purpose here is to obtain higher spatial resolution photographs of the repaired monuments, where the repair is performed by automatic detection. The resulting images and videos can then serve as an input for 3D surface estimation and eventually for creating immersive walkthrough systems, justifying the title of this book.

The book is addressed to an audience including, but not limited to, practitioners and researchers having interest in digital heritage reconstruction from a computer vision point of view. The content of this book aims to foster new research ideas in this area. All the works discussed in this book have been covered with sufficient detail and we have also illustrated these with a large number of figures for better understanding. These works are a part of the project *Indian Digital Heritage (IDH) – Hampi* sponsored by the Department of Science and Technology (DST), Govt. of India (Grant No: NRDMS/11/1586/2009/Phase-II) for which Manjunath V. Joshi has been the principal investigator. Under his supervision, the work for Chapters 2–3 was carried out by Nilay L. Khatri while working as a junior research fellow in the IDH – Hampi project at DA-IICT. The work for Chapters 4–9 has been carried out by Milind G. Padalkar as a part of his Ph.D. thesis.

Milind G. Padalkar, Manjunath V. Joshi, and Nilay L. Khatri
November 2016

# Acknowledgments

We owe our gratitude to many individuals for helping us in finishing this manuscript. We would like to use this space to express our sincere thanks to all these individuals in no particular order.

We have immensely benefited from the comments and suggestions received form the following people: Prof. S. C. Sahasrabudhe, Prof. R. Nagaraj, Prof. Anish Mathuria, Prof. Suman Mitra, Prof. Asim Banerjee, Prof. Hemant Patil, Prof. Manish Narwaria. We are also thankful to other faculty members, Mukesh Thaker and other administrative staff who have helped us directly or indirectly. We are greatly indebted to Prof. Mukesh Zaveri at SVNIT, Surat, India, and Prof. Mehul Raval at Ahmedabad University, India, for many helpful discussions toward building up the content of Chapters 4, 7 and 8. We thank Prof. Toshiyuki Amano at Wakayama University, Japan, for his valuable communication over emails and generosity in sharing the code of his work in [3].

We wish to gratefully acknowledge partial financial assistance received under the following project grant: *Indian Digital Heritage (IDH) - Hampi* under the Department of Science and Technology (DST), Govt. of India (Grant No: NRDMS/11/1586/2009/Phase-II). We are also grateful to the other groups involved in the IDH project for their support and encouragement.

We would like to thank the editor of this series Prof. Brian A. Barsky, who saw potential in our work that could be shared with a wider reader base, and encouraged us to pursue it diligently. Our sincere thanks to Michael B. Morgan for providing us with an opportunity to present our work to the Vision-Graphics fraternity and the staff of Morgan & Claypool for their very enthusiastic cooperation and assistance. They have been extremely supportive and encouraging during the whole process. We also express our heartfelt thanks to all the anonymous reviewers for their constructive suggestions that have greatly helped in improving this manuscript toward bringing it to its present form.

The first author would like to express his gratitude to Dr. Chintan Parmar, Manali Vora, Dr. Jignesh Bhatt, Dr. Rakesh Patel, Dr. Kishor Upla, Dr. Prakash Gajjar, Shrishail Gajbhar, Vandana Ravindran, Ramnaresh Vangala and Naman Turakhia for many fruitful discussions. He is also thankful to all his friends at SVNIT, Surat for their constant support and encouragement. The first author is indebted to his mother Suniti and father Gajanan for their solicitude, patience and providing constant backing and motivation.

The second author wishes to express his deep sense of gratitude to his family members Smita, Nidhi and Ninad for their constant love, encouragement and patience. He is grateful to his mentors Prof. K. V. V. Murthy, Prof. S. Chaudhuri and Prof. P. G. Poonanacha for their inspiration and support.

The third author would like to thank his mother Niharika for her untiring support and encouragement without which it would have been impossible for him to sit and burn the midnight oil to finish Chapters 2–3 after a tiring day at the office. He would like to thank his younger brother Prerak whose happiness in the author's work only added fuel to his mother's encouragement. He would also like to thank all the other loved ones (too many to name here) whose presence around him made writing an enjoyable experience rather than drudgery. He would also like to thank Bajaru who prodded, pushed, guided and inspired him to bring this work to a closure.

Milind G. Padalkar, Manjunath V. Joshi, and Nilay L. Khatri
November 2016

# CHAPTER 1

# Introduction

Historic monuments and heritage sites across the world are important sources of knowledge, which introduce us to our cultural history, depicting the evolution of humankind. They not only represent irreplaceable assets that signify the culture and civilization of the past, but also portray masterpieces of accomplishments that symbolize the human potential. Such places serve as excellent tourist attractions and can contribute significantly to a nation's gross domestic product, as tourism is of great economic importance to many industries. It is for this reason many organizations and government agencies globally have been taking keen interest toward safeguarding and preserving the heritage sites.

Over the centuries, the heritage sites have witnessed a number of natural calamities and sabotage resulting in their present ruined condition. A not very distant example is that of the infamous destruction of *the Buddhas of Bamiyan* in Afghanistan by the Taliban, as reported in Behzad and Qarizadah [7], Grün et al. [46]. Fearing the risk of further damage by visitors, access to many heritage sites is now restricted. One such example is that of the *mandapa with musical pillars* in the Vithala temple at Hampi in India, where the visitors are not allowed to touch and experience the chimes resounding from the musical pillars. In order to preserve the heritage sites, one may think of their physical renovation. However, the renovation may not only pose danger to the undamaged monuments, but may also fail to mimic the skillful historic work. Alternatively, it would be interesting to digitally reconstruct the heritage sites since such a process avoids physical contact with the monuments. The digitally reconstructed heritage site in the form of an immersive walkthrough system may then provide unrestricted access for viewing the monuments in their entirety. Also, in today's world, with the availability of better computing and storage facilities, preservation of the digitally reconstructed monuments is inexpensive. The digital reconstruction of the heritage sites also facilitates the creation of virtual tours, immersive walkthroughs and mixed-reality systems using digitized 3D models. The *Virtual Asukakyo* project [116] is one such example which digitally restored the ancient capital of Asuka-Kyo in Japan and provided a mixed-reality experience to the visitors.

Reconstruction and creation of immersive walkthrough systems using digitized 3D models involves the processing of input captured using various acquisition systems viz. laser scanners, multiple photographs for dense reconstruction, etc. Since laser scanners are expensive and usually provide only depth information, one can estimate the same with the help of a large number of photographs wherein color information is already available, using the technique proposed in Furukawa and Ponce [37]. However, in either of the acquisition systems, greater details may not be captured, and self-occlusion or difficulty in capturing the scene/object from a particular viewpoint may exist. This results in missing the high-resolution (HR) details as well as leaving holes in the captured/estimated 3D surface. The holes can be filled up using the Poisson surface reconstruction method proposed in Kazhdan et al. [60]. However, in this case, the reconstructed surface is over-smoothened and again the HR details are lost.

One can overcome the above problem by performing resolution enhancement of the captured photographs before estimating the 3D surface. Also, filling of the missing regions in the photographs can be performed before estimating the depth values. While high-resolution details can be obtained using super-resolution, inpainting can be used to digitally fill or repair the entire damaged region in the photographed scene. Particularly, when heritage monuments are to be digitally preserved, one needs to bear in mind the excitement of the viewers for observing fine details of the skillful historic work and their desire to view the monuments in an undamaged form. By super-resolving and digitally correcting any damaged regions by means of inpainting, the viewers can be provided with an enhanced visual experience. Thus, the creation of immersive walkthrough systems or digital reconstruction of invaluable artwork consists of image super-resolution and inpainting as the preliminary steps.

This book presents methods for image super-resolution and techniques for automatically detecting and inpainting regions like cracks and other damaged regions in heritage monuments. The purpose here is to obtain higher spatial resolution photographs of the repaired monuments, where the repair is performed by automatic detection. The resulting images and videos can then serve as an input for 3D surface estimation and eventually for creating immersive walkthrough systems. This justifies the title of this book. In what follows, we provide a quick introduction to super-resolution and inpainting followed by a summary of the content.

## 1.1   WHAT IS SUPER-RESOLUTION?

A digital image is generated by spatially sampling the continuous scene acquired using an image sensor. If the sampling frequency is low, it introduces distortion due to aliasing in the high frequency components. In addition, during image acquisition the optical distortions are also introduced due to lack of focus, lens aberration, diffraction limit, etc. The sensor *Point Spread Function (PSF)*, relative motion between the camera and original scene, or shutter speed introduce motion blur in the captured image. The insufficient sensor density and various electronics components in the system also contribute noise to the final digital image. So, image super-resolution (SR) refers to an algorithmic approach to reconstruct a high spatial resolution image from one or more low-resolution (LR) observations affected with the aforementioned distortions. In effect, the SR process attempts to minimize aliasing and various imaging distortions. Figure 1.1 illustrates the image formation model and a usual HR reconstruction process for multi-image methods.

The existing cue-based super-resolution methods for HR reconstruction can be classified into two categories: *motion-based* and *motion-free* techniques. Motion-based techniques use the relative motion between multiple LR observations as a cue to estimate the high-resolution image.

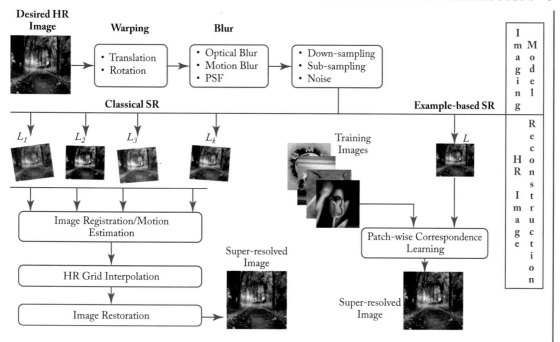

**Figure 1.1:** A bird's eye view of super-resolution: A typical flow to reconstruct super-resolved images using multi-image and example-based SR techniques.

These methods are widely known as *classical* SR, while motion-free techniques use cues such as blur, zoom, and shading. The latter methods, despite requiring multiple observations, do not use the relative motion between these observations. A comparative survey of the motion-based and motion-free SR techniques can be found in the works by Chaudhuri [16], Chaudhuri and Joshi [17], Park et al. [92]. Some researchers have also attempted to solve the super-resolution reconstruction problem without considering any specific cue, and instead use an ensemble of images as a training set in order to learn the required information for resolution enhancement. These are popularly known as *example-based* SR techniques in which the relationships between LR-HR patches are learned from a database containing pairs of LR-HR images or from the given image itself. The majority of the super-resolution methods proposed in last few years fall in *example-based* SR techniques. A comprehensive survey of both classical SR and example-based SR, and their various sub-categories along with the problem domains they are applied to, is provided in Nasrollahi and Moeslund [84].

The multiple LR images of a scene are basically different "looks" of the same scene from different view points, i.e., they are sub-sampled (aliased) and sub-pixel shifted versions of a reference HR image. An integer shift unit between images provides no new information, and hence such LR images cannot be used to reconstruct an HR image. However, if the LR images have different sub-pixel shifts from each other, then each LR image contains information that cannot be found in other LR images, and so can be exploited to obtain an HR image. These different looks of a scene can be obtained by a relative motion between the camera and the scene. Multiple scenes can be obtained by moving a single camera or from multiple cameras located in different positions. If these scene motions are known or can be estimated within sub-pixel accuracy, then this information can be combined with LR images to reconstruct an SR image. Although the details may vary, a typical classical SR algorithm involves three stages (see Figure 1.1): image registration (also known as motion estimation), HR grid interpolation, and image restoration (deals with deblurring and noise removal). The image registration stage estimates the relative shifts between LR images with reference to one of the LR images (with a fractional pixel accuracy). The accurate sub-pixel motion estimation is a very important factor in the success of any motion-based (classical) SR image reconstruction algorithm. Since the shifts between LR images are arbitrary, the registered HR image does not always match up to a uniformly spaced HR grid. Thus, non-uniform interpolation is necessary to obtain a uniformly organized HR image from a non-uniformly organized composite of LR images. Finally, image restoration is applied to this up-sampled image to remove blurring and noise. The entire process of LR image formation from an HR image can be expressed in the form of a mathematical model as:

$$\mathbf{y}_k = \mathbf{D}\mathbf{B_k}\mathbf{M_k}\mathbf{x} + \mathbf{n}_k \quad \text{for } 1 \leq k \leq n, \tag{1.1}$$

where, $\mathbf{x} = [x_1, x_2, ..., x_N]^T$ is a lexicographically ordered HR image; $\mathbf{y_k} = [y_{k,1}, y_{k,2}, ..., y_{k,M}]^T$ denote a lexicographically ordered $k^{th}$ LR image; $M = N_1 \times N_2$ is the resolution of each LR image. $\mathbf{M_k}$ is a warp matrix (rotation, translation etc.); $\mathbf{B_k}$ represents a blur matrix and $\mathbf{n_k}$ is a lexicographically ordered noise vector. The readers are encouraged to refer Park et al. [92] for a detailed discussion of this model along with an exhaustive and excellent review of various classical SR algorithms. The literature of classical SR is rather rich, and it is beyond the scope to discuss all of them here; however, some very important works in this area deserve a mention. The classical SR approach was first addressed in the work by Huang and Tsai [54] which demonstrated the estimation of an HR image using a number of LR observations with sub-pixel shifts. Later, Irani and Peleg [55] presented an iterative back-projection-based approach in which the super-resolved image was estimated by iteratively computing the difference between the observed and the simulated LR images from the current SR estimate. An alternate approach which uses $l_1$-norm minimization and a robust bilateral prior was proposed much later by Farsiu et al. [31]. Figures 1.2b–c demonstrate an example of classical SR for the magnification factors of 2 and 4.

The classical SR methods have had their own fair share in the limelight. However, by this time an upper limit on the achievable magnification factor with the classical SR was proven to be around $\approx 2$ in the works of Baker and Kanade [5], Lin and Shum [70]. This limitation motivated the exploration of alternative techniques to attain higher magnification factors for super-resolution. Because the richness of real-world images is difficult to capture analytically, researchers have been exploring learning-based approaches for enlarging images for a couple of decades now.

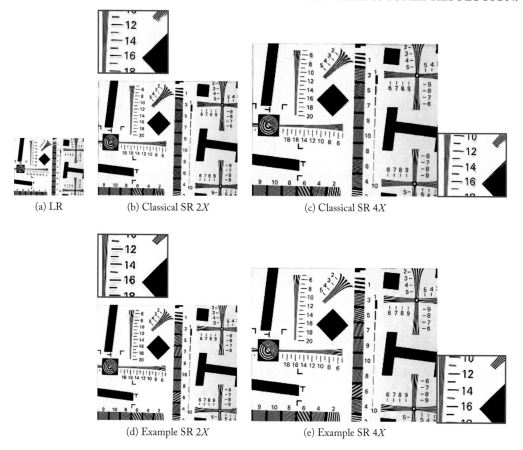

(a) LR  (b) Classical SR 2$X$  (c) Classical SR 4$X$

(d) Example SR 2$X$  (e) Example SR 4$X$

**Figure 1.2:** A demonstration of classical SR and example-based SR: As can be observed in (e) and (c), the perfomance of classical SR is inferior to that of example-based SR for a magnification factor greater than 2. The results in (b) and (c) are obtained using four LR images. Due to space constraints, the training images used for example-based SR images are not included here. The training set comprised four images primarily containing text. The insets show the performance of both types of SR methods for two different magnification factors. Image courtesy of Vandewalle et al. [123].

In a training set, the algorithm learns the fine details that correspond to different image regions seen at a low-resolution and then uses those learned relationships to predict fine details in other images (see Figure 1.1). These techniques are often referred to as "example-based" methods. Freeman et al. [36] were the first to propose such an example-based method for SR. This approach required the construction of an over-complete dictionary consisting of LR-HR patch-pairs obtained from a large set of different training images. Given the test LR image, the generated dictionary was then used to recover the missing HR details. By placing sparsity constraints on the over-complete dictionaries Yang et al. [135] proposed a technique using the concept of compressive sensing in which the SR problem was solved by jointly training LR-HR pair dictionaries and enforcing the similarity of the corresponding sparse representations. Figures 1.2d–e illustrate an example of example-based SR for the magnification factors of 2 and 4.

Many super-resolution approaches use some form of smoothness prior that has the tendency to smooth out the textured regions of images for high magnification factors. In order to preserve edges in the super-resolved image, Fattal [32] proposed an approach that used edge dependencies between different resolution versions of an image. Likewise, a gradient profile prior was used in the work by Sun et al. [114], which parametrically defines the shape and sharpness of image edges learned from a large number of natural images. Gajjar and Joshi [39] proposed a wavelet learning-based SR approach by making use of a detail-preserving *Inhomogeneous Gaussian Markov Random Field* (IGMRF) prior. Here the learning process provides an initial estimate which is used in estimating the super-resolved image. Later, Freedman and Fattal [35] proposed an approach that uses a non-dyadic filter to preserve the HR image details.

Motivated by the self-similarity properties of images across scales studied by Ruderman and Bialek [104], Turiel et al. [122], researchers also exploited the patch repetitions that occur within and across different resolution scales of the given LR image in order to achieve super-resolution. A unified framework combining the classical and example-based SR was proposed by Glasner et al. [42], in which similar LR patches were found within and across different image resolution scales. Here each similar patch to the LR patch imposes a constraint on the corresponding HR patch which is estimated by solving a system of linear equations arising from these constraints. A similar approach was proposed by Luong et al. [76] which uses kernel regression to fuse the similar patches in order to obtain super-resolved image. The most recent approaches for example-based SR make use of anchored neighborhoods along with learned offline-regressors [119] and transformed self-exemplars [53] which allow geometric variations of the example patches in order to cover significant numbers of textural appearance variations in the scene.

Recently, deep learning-based approaches have provided state-of-the-art results in many areas, including image super-resolution. These methods [22, 26, 27, 71, 130] learn a cascade of filters and use them to enhance the spatial resolution of an input image. It is interesting to note that a sparsity prior is used in most of these methods to learn the cascade of filters.

## 1.2   WHAT IS INPAINTING?

Image inpainting is the process of restoring or modifying the image contents imperceptibly. Given an image and a region of interest in it, the task of an inpainting process is to fill up the pixels in this region, in such a way that either the original content is restored or the region is visually plausible in the context of the image. An example of inpainting is shown in Figure 1.3. Here one can see that although the sign-board marked in Figure 1.3b is not restored, the inpainted image shown in Figure 1.3c has the marked region filled in such a manner that it is difficult to comment about which region is synthesized.

(a)                                          (b)                                          (c)

**Figure 1.3:** An example of inpainting. (a) Input image; (b) region of interest to be inpainted supplied by the user is shown by the white region with red boundary; (c) result of inpainting. Image courtesy of Criminisi et al. [21].

Image inpainting can be used for a number of applications that require automatic restoration or retouching of some region of a photograph. In fact, the term "inpainting" is derived from the art of restoring damaged images in museums by professional restorers [8]. Although restoration and inpainting are used interchangeably, inpainting can be considered as a superset of restoration. In general, restoration refers to the undoing of degradation, while inpainting also allows creating special effects such as removing/adding objects.

Image inpainting has been an active area of research for the past decade. During the 1990s many researchers addressed the problem of interpolating missing pixel values, but it was only toward the end of the decade that Masnou and Morel [80] proposed the first inpainting technique based on partial differential equations (PDEs). Their method connected the level lines (i.e., curves or contours of constant intensity) arriving at the boundary of an occluded region. With this technique the occluded regions were filled but the level lines did not curve in a plausible manner. This is illustrated with an example shown in Figure 1.4. A major breakthrough to the inpainting problem was later provided by Bertalmio et al. [8] by proposing another PDE-based technique. Their algorithm not only connected the level lines arriving at the boundary of the occluded regions, but also enabled their curving inside the occluded region in a visually plausible manner. The periodical curving of the level lines that also avoids crossings was achieved using anisotropic diffusion [2, 98]. The algorithm proposed by Bertalmio et al. [8] was successful in propagating structure, however, failed in inpainting areas with large textured regions. Nevertheless, this method inspired future works that also relied on the propagation of the level lines [86, 129].

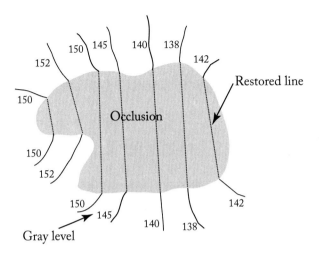

**Figure 1.4:** An example of connecting level lines across an occlusion. Image courtesy of Masnou and Morel [80].

For propagating texture, a patch-replicating method was suggested by Criminisi et al. [20, 21]. This patch-based technique exploits the self-similarity in an image by searching for a similar patch from the surrounding known regions (having no missing pixels) to inpaint the missing pixels in a patch under consideration. Here the occluded region containing the missing pixels is filled in a patch-by-patch approach by copying pixels from the corresponding similar patch, i.e., exemplar. The method emphasizes the order of selecting the patch to be filled, since it allows the propagation of both structure as well as texture. Figure 1.5 shows an example of inpainting performed using the PDE-based method proposed by Bertalmio et al. [8] (which appears blurred) in comparison to the patch-based technique proposed by Criminisi et al. [21] (showing a more convincing filling of the missing region).

Another technique to inpaint large textured regions was proposed by Pérez et al. [96], Pérez et al. [97], that makes use of Poisson's equation for adding objects/texture from other images by using them as a guidance vector field. Thus, if a user supplies the region of interest to be edited and a region from which information is to be transferred, i.e., a guidance vector field, the technique results in a seamless blending. An example of the inpainting performed using this method is illustrated in Figure 1.6. Researchers have also proposed methods [45, 132] that make use of level lines for texture synthesis.

We observe that the missing pixels can be filled either by gradually propagating information from outside the boundary of the region of interest (ROI) or by making use of cues from similar patches. Based on these filling strategies, the existing inpainting methods can be categorized

(a)                    (b)                    (c)                    (d)

**Figure 1.5:** Inpainting a missing region using PDE- and patch-based methods. (a) Input image; (b) region of interest to be inpainted supplied by the user is shown by the white region with red boundary; (c) inpainted result using the PDE-based technique proposed in Bertalmio et al. [8]; (d) inpainting by propagation of texture using the patch-based approach proposed in Criminisi et al. [21]. Image courtesy of Criminisi et al. [21].

into two important groups viz. methods using level lines and solving partial differential equations (PDEs) [8, 80] and those based on exemplars [20, 21, 97, 131]. Also, inpainting techniques based on probabilistic structure estimation [110], methods using depth and focus [81], etc., have been influential. Recent techniques include the works by Ghorai et al. [41], Huang et al. [51], Purkait and Chanda [100]. While Purkait and Chanda [100] present a patch-based anisotropic diffusion technique to generate high-frequency details for enhancing line/brush strokes usually found in mural paintings, the method by Ghorai et al. [41] performs multiscale image completion by combining transform domain patch filtering. Likewise, Huang et al. [51] propose a patch-based image-completion method that makes use of constraints on patch offsets and transformations derived from the translational regularities in the estimated planes. However, all these methods are semi-automatic, in the sense that regions to be inpainted are required to be manually selected by the users.

*Blind inpainting* is one of the categories of inpainting techniques that do not require any user-interaction for providing the regions to be inpainted. In fact, these techniques perform inpainting without *a priori* knowledge or detection of the missing pixels. However, they assume the input image to be a noisy observation and perform image recovery by considering the pixels to be corrupted by various noise or degradation models [1, 25, 134]. Xie et al. [133] and Cai et al. [10] have used deep neural networks for blind inpainting, wherein the neural networks are trained using a large number of examples of corresponding corrupted and uncorrupted patches. Figure 1.7 illustrates a result obtained using the blind inpainting method proposed by Cai et al. [10]. The blind inpainting methods are used when there is difficulty in selecting the corrupted

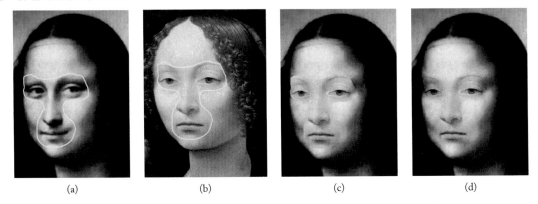

(a)          (b)          (c)          (d)

**Figure 1.6:** Illustration of inpainting using the Poisson image editing method [97]. (a) Input (desti-nation) image showing the region to be inpainted inside the region marked by the yellow boundary; (b) image containing the source region marked by a yellow boundary; (c) copying of the source region inside the destination image, i.e., cloning; (d) seamless cloning, i.e., inpainting the destination image using the source region as the guidance vector field. Image courtesy of Pérez et al. [97].

(a)                              (b)

**Figure 1.7:** Illustration of blind inpainting. (a) Input image; (b) blind inpainted image, i.e., inpainting is performed without any knowledge about which pixels in (a) are missing. Image courtesy of Cai et al. [10].

(i.e., missing) pixels. One may note that the blind inpainting techniques perform well when the missing pixels are not localized but randomly spread across the complete image. Moreover, the inpainted regions appear blurred and they fail to recover the texture in large regions. Another problem here is that the known pixel values also get modified due to the blind nature of these inpainting techniques.

A different category of techniques that also facilitate the automatic detection of the re-gions to be inpainted is known as *auto-inpainting*. Unlike those techniques that are supplied with the regions to be inpainted, the amount of published literature under this category is very much limited. One may note that techniques under this category are developed to perform automatic detection of regions to be inpainted for specialized applications and may or may not follow with an implicit method for inpainting the detected regions. Once the regions are automatically detected, the missing pixels in these regions can be filled using a generic inpainting technique. Thus, such methods can avoid the need of manually supplying the regions to be inpainted. Also, unlike blind inpainting, the advantage here is that these techniques fill only the detected missing pixels without modifying values of the known pixels. Our methods discussed in this book for digital repair of damaged monuments fall under the category of inpainting and auto-inpainting techniques. Following is a brief overview of techniques under the auto-inpainting category.

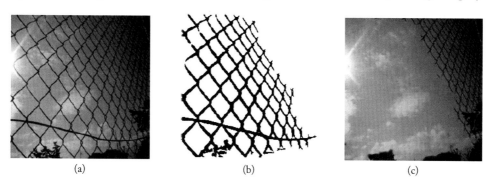

(a)         (b)         (c)

**Figure 1.8:** Illustration of auto-inpainting. (a) Input image; (b) automatically detected pixels to be inpainted; (c) inpainted image obtained by filling the pixels detected in (b). Image courtesy of Tamaki et al. [115].

A method to detect damage in images due to color ink spray and scratch drawing was proposed by Chang et al. [15]. Their method makes use of several filters and structural information of damages. Tamaki et al. [115] address the detection of visually less important string-like objects that block the user's view of a discernible scene. The method requires the occluding objects to have high contrast with respect to the background and is restricted to those objects that are long and narrow. Figure 1.8 illustrates an auto-inpainted result obtained using this method. Amano [3] presented a correlation-based method for detecting defects in images. This method relies on the correlation between adjacent patches for detection of defects, i.e., a small number of regions disobeying an "image description rule" complied by most local regions. The method works well for detecting computer-generated superimposed characters having a uniform pattern. Parmar et al. [93] proposed a technique which uses matching of edge-based features with pre-existing templates to distinguish vandalized and non-vandalized regions in frontal face images of monuments at heritage sites. Similarly, Turakhia et al. [121] proposed a method to automatically inpaint cracks in images of heritage monuments which relies on the detection of edges and tensor voting to identify cracks. For the detection of cracks in pavement images Zou et al. [140]

proposed a method that also makes use of tensor voting and is heavily dependent on the accuracy of generating the crack-pixel binary map that acts as an input to the tensor voting framework. Recently, Cornelis et al. [19] have proposed a method for virtual restoration of paintings. The method is flexible as the user may set parameters to suit the input. However, it is suitable only for the detection of fine cracks that appear in paintings. An example of a result obtained using this method is shown in Figure 1.9.

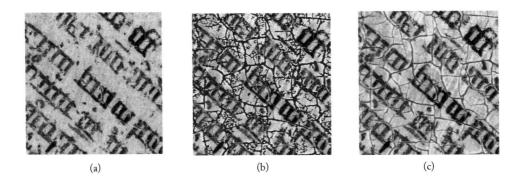

(a)                         (b)                         (c)

**Figure 1.9:** Example of auto-inpainting fine cracks appearing on paintings. (a) Input image; (b) automatically detected cracks to be inpainted; (c) inpainted version obtained by removal of the cracks. Image courtesy of Cornelis et al. [19].

For inpainting in videos, a method has been proposed by Patwardhan et al. [95]. Their technique considers a static background with a moving foreground, any of which could fall under the region to be inpainted. First, the occluded foreground patches are filled up using motion-inpainting. The background patches which are visible in other frames are then directly copied. Finally, any missing region is filled up using the exemplar-based inpainting approach [21]. It may be noted that in this approach, the users need to manually specify the objects or regions that are to be inpainted. Also, many constraints are placed on the camera motion. While a detailed and comparative survey on video inpainting techniques can be found in the works by Ghorai et al. [40], Newson et al. [85], Shih et al. [111], one may note the dependency of these techniques in the robustness of the tracking algorithms used, the lighting conditions and the constraints on camera motion. Most video inpainting techniques can inpaint moving objects under constrained camera motion or a user-selected object. However, to the best of our knowledge, there does not exist any approach that demonstrates video inpainting under unconstrained camera motion with no moving objects and is completely automatic without the need to provide the regions to be inpainted, apart from the one we discuss later in Chapter 8.

# 1.3 APPLYING SUPER-RESOLUTION AND INPAINTING IN DIGITAL HERITAGE IMAGES: CHALLENGES AND SOLUTIONS

Having discussed the current research status in super-resolution and inpainting, we now highlight the challenges in heritage reconstruction and our solutions by applying proposed super-resolution and inpainting techniques. For reconstruction, the heritage sites are first required to be digitized using a suitable acquisition method. The book by MacDonald [78] describes the process of digitalization for cultural heritage and provides information about the required hardware and software setup, acquisition of images for 2D and 3D rendering, along with related case studies. Likewise, the preservation of original objects at the heritage site or in museums as well as creating their imaged replica in digital form is discussed by Munshi and Chaudhuri [83], where the authors also discuss the issues involved in digital content and community building for archiving and global sharing of heritage resources. However, due to the limitation of capturing environments, the major challenges faced are: (a) the acquisition of high-resolution samples, which requires very expensive scanners and (b) recovery of missing regions that are generated due to difficulty in capturing from certain viewpoints. Moreover, even if the acquired data contains no missing regions, the scanned objects themselves could be damaged, leading to holes in the acquired data.

In this book, we discuss the above mentioned aspect of digitizing the cultural heritage viz. reconstruction and recovery of details in the lost or deteriorated regions in the photographs of the monuments, which is not addressed in any of the books listed above. We discuss the super-resolution and auto-inpainting methods that address the issues specific to digital reconstruction of heritage monuments which may be damaged due to weathering, natural calamities or vandalization. Using our proposed super-resolution methods one can obtain high-resolution details of the monuments, and our proposed auto-inpainting methods automate the detection of cracks or damages and perform their repair in a seamless manner. One can then use the resulting images and videos as an input for 3D surface estimation and eventually for creating immersive walkthrough systems.

The existing techniques [42, 75, 135] super-resolve natural images by estimating the HR patches from the given LR image and its multiple coarser resolutions obtained using downsampling operation. The HR patches are found in the given resolution or in the coarser resolutions. Unlike these methods, ours is a different approach in which we consider only two resolutions, viz. the given LR image and its coarser resolution version that are captured by the same camera. In order to get the HR patch, we exploit the recurrence of visual content in the given image itself, which is a typical characteristic of heritage scene images. This is done by making use of the coarser resolution version of the LR image. This self-learning approach using a captured pair of images enables one to mimic the skillful work in historic sculptures and carvings at higher magnification factors. Our second method for super-resolution builds on the first approach and provides a smart and computationally efficient technique for higher image-upsampling factors.

Higher magnification results in better details that can be provided to an immersive walk-through system.

In this book, we also propose two novel inpainting techniques based on the user-supplied regions to be inpainted for the recovery of missing regions, wherein we make use of the artistic details available in the given image itself. By doing so, we are able to imitate the creative expressions of the artists which could be in the form of brush strokes in mural paintings or carvings in petroglyphs and sculptures. One of our proposed methods not only inpaints the given image but also creates the missing details in its higher resolution, i.e., super-resolution inpainting. This is particularly suited for digitizing the artwork since both the super-resolution and the inpainting are performed simultaneously by the proposed method. Recently Ghorai et al. [41], Purkait and Chanda [100], have proposed inpainting methods for repair of heritage sites. Purkait and Chanda [100] present a patch-based anisotropic diffusion technique to generate high-frequency details for enhancing line/brush strokes usually found in mural paintings. Although this method repairs the imaged murals, it requires a skilled user to interact with the system for identifying the damaged regions. Likewise, the method in [41] performs multiscale image completion by combining transform domain patch filtering. In these methods, the detection of cracks is performed in a user-assisted interactive manner and may require special imaging conditions. On the other hand, our proposed inpainting methods require no special imaging conditions.

For the recovery of missing regions, another issue is to identify which regions in the photographed scene are to be repaired. While human beings usually have a consensus in identifying defaced regions and cracks as the damaged regions that need to be repaired, automation of this task is a major challenge. By doing so, one can fully automate the inpainting process, i.e., inpainting can be performed without requiring user interaction. Algorithms to digitally detect and restore typical damages that photographs suffer, such as foxing, water blotches, fading and glass cracks, are discussed in Stanco et al. [113]. These methods aim to undo any damage to the photograph itself rather than digitally repairing the imaged physical object. However, the digital repair of heritage sites requires the detection of defects in the objects being photographed.

The defect-detection method proposed by Amano [3] works well for detecting computer-generated superimposed characters that have a uniform pattern. Also, Zou et al. [140] have proposed a method for detecting defects in the form of pavement crack. Their method makes use of tensor voting, but it is, however, heavily dependent on the accuracy of generation of a crack-pixel binary map. Cornelis et al. [19] have proposed a method for detection of cracks. Nevertheless, their method is suitable only for the detection of fine cracks that appear on paintings and requires fine-tuning of many parameters for proper detection. In this book, we discuss methods particularly tailored for automatic detection of damaged regions in facial images of statues and cracks in heritage scenes, which to the best of our knowledge have not been attempted previously. Unlike other approaches, our proposed methods detect the defect in the physical object being photographed and not in the photograph itself. Our first method detects the visually dominant eye, nose and lip regions in statues and then identifies which of these are damaged. We next discuss techniques for detecting cracks, and the detected regions are inpainted to obtain the reconstructed view of the photographed monument/scene.

We also extend our work on crack detection to perform auto-inpainting in videos. The video inpainting method proposed in Patwardhan et al. [95] requires the users to manually specify the objects or regions that are to be inpainted. Also, many constraints are placed on the camera movement. Our proposed method for video-inpainting can detect and inpaint cracks without constraining the movement of the camera. Thus the method is completely automated and needs no human interaction. This can be particularly useful for performing on-the-fly digital reconstruction of damaged regions when tourists capture the heritage monuments using their handheld video-capturing devices.

## 1.4    A TOUR OF THE BOOK

In Chapter 2 we discuss an image super-resolution method based on self-learning and Gabor prior. Here the idea is to exploit the recurrence of visual content in the given image to mimic the skillful historic work at higher magnification factors. This is done by constructing dictionaries of low- and high-resolution patch pairs from the given image and its single coarser resolution. A high-resolution image is estimated by making use of these dictionaries, i.e., self-learn the HR patches corresponding to the LR patch in the given image. The estimated high-resolution image is then regularized using the Gabor prior as this preserves various levels of detail, resulting in better super-resolved image.

The discussion of a super-resolution approach using self-learning in Chapter 2 uses the given LR image to achieve high magnification factors by constructing LR-HR dictionaries. However, when we attempt super-resolution for higher magnification factors, the dictionary sizes dramatically increase, and make it computationally prohibitive. In order to overcome this difficulty, Chapter 3 presents a method of super-resolution that removes the redundancies inherent in large self-learned dictionaries and also avoids the use of any regularization that drastically reduces time complexity. The method uses variance of pixel values in patches as a cue to get rid of the redundancy.

After discussing techniques to obtain super-resolution with high magnification factors, we propose an inpainting technique for filling up the missing regions in the object/scene being imaged, in Chapter 4. In order to imitate the creative expressions of the artists, which could be in the form of brush strokes in mural paintings or carvings in petroglyphs and sculptures, one needs to capture the dependencies in the regions using pixel-neighborhood relationships. In this chapter, we discuss an iterative exemplar-based inpainting technique wherein we estimate the parameters of an autoregressive (AR) model to represent the pixel-neighborhood relationships. Unlike those approaches which simply copy the pixels to be inpainted from the best matching exemplar, we use the AR parameters in addition to the best matching exemplar to fill the missing pixels so that the artistic creative expressions are retained in the filled regions.

Chapter 5 provides a discussion on methods that we attempted toward improving inpainting. Our observations and the conclusion drawn from these attempts were instrumental in the development of our technique for simultaneous inpainting and super-resolution. In a quest to find a more reliable source exemplar for inpainting, the first approach in Chapter 5 finds a match of a larger window around the patch to be inpainted, and a refinement is attempted by copying pixels from a corresponding patch in the matched window. In the second approach for refinement, a relation of a patch with its first-order neighbors obtained in the least-squares sense was used. The third refinement approach considered the relationship of a patch with other patches within a window as a sparse representation using the compressive sensing framework. A different approach for inpainting based on curvature was also attempted. Here an optimal value of the pixel at the boundary of the region to be inpainted is estimated by minimizing the difference in curvature value at adjacent pixels. Though these approaches did not result in better inpainting, they motivated us to use the idea of dictionary-based learning and compressive sensing for finding reliable exemplars to obtain better inpainting results as discussed in Chapter 6.

As already discussed, for applications like creating immersive walkthrough systems or digital reconstruction of invaluable artwork both super-resolution and inpainting of the given images are the preliminary steps that need to be performed for providing a better visual experience. In Chapter 6, we present a unified framework to perform simultaneous inpainting and super-resolution, rather than addressing them in a pipelined manner as is usually done. The presented approach is based on the self-learning concept introduced in Chapter 2. With the use of this technique, the missing pixels are filled not only in the given spatial resolution but also in the higher resolution leading to super-resolution inpainting.

In practice, inpainting techniques require the user to manually select the regions to be inpainted. To this end, in the next two chapters, we discuss auto-inpainting techniques for certain applications that do not require specifying of the region of interest. In other words, the missing region to be filled up is not known but is automatically detected prior to inpainting. In Chapter 7, we discuss a method that automates the process of identifying the damaged dominant regions viz. eyes, nose and lips in the face images of statues at a historic site, for the purpose of inpainting. The dominant regions are extracted by considering the bilateral symmetry of a face. Texton features are then used to identify the damaged regions, based on existing templates of damaged and non-damaged regions. The Poisson image editing method is then used to inpaint the damaged regions.

The aim of Chapter 8 is to introduce techniques that can automatically detect the cracked regions in heritage monuments and demonstrate their repair by inpainting. This can be particularly useful for performing on-the-fly digital reconstruction of damaged regions when tourists capture the heritage monuments using their handheld video-capturing devices. We first discuss a simple method to automatically detect the damaged regions which are characterized by abruptly dark deteriorations in the photographed monuments of a heritage site by making use of order-statistics and density filters. This is followed by a singular value decomposition (SVD)-based technique for automatic detection of the cracked regions in the photographed object/scene, for the purpose of digitally restoring them to their entirety using inpainting. Further, we discuss another effective and more accurate crack-detection method based on the comparison of patches using a measure derived from the edit distance, which is a popular string metric used in the area of text mining. We extend this crack-detection approach to perform inpainting of video frames by making use of the scale invariant feature transform (SIFT) and homography. Here we consider the camera movement to be unconstrained while capturing video of the heritage site, since such videos are typically captured by novices, hobbyists and tourists. Finally, we provide the temporal consistency measure to quantify the quality of the inpainted video.

Chapter 9 provides brief insights on challenges and future directions in super-resolution and inpainting for heritage reconstruction. The aim of this chapter is to motivate the readers toward pursuing these challenges in the future.

The research toward advancement in the area of digital heritage reconstruction has been constantly evolving and it is quite natural that the works presented in this book have earlier been reported in a few workshops, conferences and journals. This book derives some of its content from these publications: Khatri and Joshi [62, 63], Padalkar and Joshi [87], Padalkar et al. [88, 89, 90, 91].

# CHAPTER 2

# Image Super-resolution: Self-learning, Sparsity and Gabor Prior

Enhancement of spatial resolution is a preliminary step for digital reconstruction of heritage sites as this enables the viewers to perceive finer details of the skillful historic work. This chapter presents a super-resolution approach for resolution enhancement that exploits the recurrence of the visual content in natural images, which is an important characteristic of the images captured at heritage sites. Glasner et al. [42] have shown that for a sufficiently small-sized patch (e.g., $3 \times 3$ or $5 \times 5$), there exist many similar or same patches within and across different resolution scales of a natural image, as illustrated in Figure 2.1. Their approach for super-resolution from a single image uses this cue by constructing an LR cascade consisting of successively downsampled versions of the input LR image. For every source patch in the LR image, similar patches are searched within the LR cascade, and corresponding parent HR patches are used as priors for image super-resolution. As the approach relies on the match found for a particular patch within the LR cascade, the resolution increase depends on the size of the image. Thus, for images where spatial resolution is low and required magnification factor is high, the probability of finding a similar patch of the same size within the LR cascade is small, and subsequently such patches are simply interpolated to obtain corresponding HR patches. Since the patches that do not find any match are simply interpolated, the final super-resolved image suffers in quality.

The approach presented in this chapter makes use of the input LR image and only its single coarser resolution version captured using a real camera to achieve super-resolution with high magnification factors. It may be of interest to note that the pair of LR images and their coarser resolution version can be captured using a camera or can be obtained simply by downsampling of the LR image. However, the experimental results shown in Section 2.4 are performed on images obtained using the former method. Dictionaries consisting of the *best-mapped* LR patches from the given LR image and their corresponding HR patches, which are otherwise interpolated in Glasner et al. [42], are used to estimate the missing HR patches. This avoids the aliasing effect that may result due to interpolation. The HR patches are estimated by incorporating dictionary learning and compressive sensing into the SR framework. The LR-HR patch pairs are used to estimate the degradation matrix, and the final super-resolution is achieved using a regularization framework in which a novel prior, called *Gabor prior*, is introduced. The Gabor filter for image

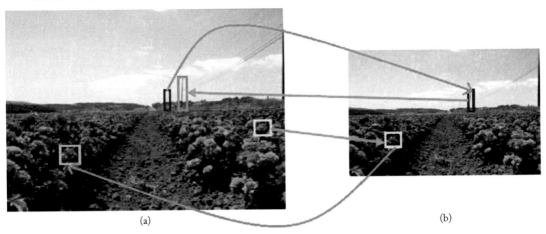

(a)                                                    (b)

Figure 2.1: **Patch recurrence property in a natural image:** (a) LR image $I_0$; (b) $I_{-1}$: the coarser resolution version of $I_0$. Note that a small patch in $I_0$ has a similar patch of the same size in $I_{-1}$ (see blue and yellow boxes) that is used to extract corresponding HR patch from $I_0$ (see green and pink box) (best viewed in color). Image courtesy of Khatri and Joshi [62].

super-resolution is employed in Lu et al. [73], where features extracted from an LR image using the Gabor filter bank are used to estimate the features in the HR image patches. Contrary to this approach, self-learning uses the Gabor prior in regularizing the solution. This preserves the details at various levels and yields a better estimate of the super-resolved image.

## 2.1   SINGLE-IMAGE SR: A UNIFIED FRAMEWORK

The super-resolution methods introduced in later sections of this chapter take their inspiration from ideas proposed in Glasner et al. [42]. The approach proposed unifies the principles of classical and example-based methods to achieve super-resolution from a single image. Patch recurrence within image is used to apply classical SR constraints, while patch recurrence across different resolution scales forms the basis for example-based SR. A patch "recurs" in another scale if it appears "as is" without any blurring, sub-sampling or scaling in a scaled down version of the image. Having found a similar patch in a smaller image-resolution scale, its HR parent can be extracted from the input image (see Figures 2.1 and 2.4). Each LR patch with its HR parent forms an LR-HR pair of patches.

When patches of very small size (e.g., $5 \times 5$) are considered, they tend to recur abundantly within an image due to the fact that very small-sized patches usually contain a single edge or corner. Moreover, distortions introduced due to perspective projection during image capture leads to scene-specific information in diminishing sizes, thus recurring in multiple scales of the image. Readers are referred to Zontak and Irani [139] for an insightful analysis and applications of the

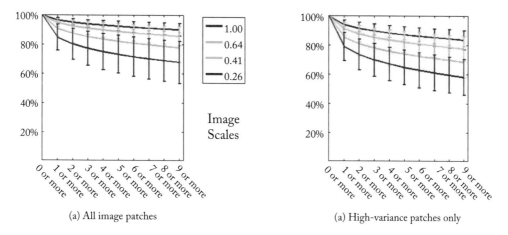

(a) All image patches                    (a) High-variance patches only

**Figure 2.2:  Average patch recurrence within and across resolution scales of an image:** (a) the percent of image patches for which there exist $n$ or more similar patches ($n = 1,2,3,...,9$); (b) the same statistics, but measured only for image patches with the highest intensity variances (top 25%). These patches correspond to patches of edges, corners and texture. Image courtesy of Glasner et al. [42]

recurrence property of image patches to various vision problems. Figure 2.2 presents the average recurrence rate of patches with different sizes across multiple resolution scales of an image. Even for high-variance patches the lowest recurrence rate is around 60%. Sections 2.1.1–2.1.3 briefly introduce the readers to the ideas used by Glasner et al. [42] to achieve super-resolution from a single LR image using patch recurrence property.

## 2.1.1    CLASSICAL (WITHIN-SCALE) SUPER-RESOLUTION

The goal of classical SR is, given a set of LR images $\{L_1, L_2, ..., L_n\}$ (at sub-pixel misalignments) of the same scene, to recover their corresponding HR image $H$. Each $L_j$ ($j = 1,2,..,n$) is obtained from $H$ by a blurring and sub-sampling process. Let us say $B_j$ is the *Point Spread Function (PSF)* for LR image $L_j$. So, each pixel $p = (x, y)$ in $L_j$ induces a linear constraint on the unknown intensity values within the local neighborhood around its corresponding pixel in $q$ in HR image $H$ ($q \in H$). When only a single LR image $L$ is available, the problem of recovering $H$ becomes under-determined as the constraints induced by $L$ are smaller than the unknowns in $H$. However, when enough LR images are available (at sub-pixel shifts), the number of constraints induced exceeds the unknowns. These constraints can be expressed as: $L_j(p) = (H * B_j) = \Sigma_{q_i \in Support(B_j)} H(q_i) B_j(q_i - q)$, where $H(q_i)$ are the unknown intensity values around pixel $q$ in HR image $H$ for pixel $p$ in $L_j$. The patch-recurrence property in an image $L$, as observed earlier, can be exploited to solve these systems of linear constraints to estimate $H$. Let $p$ be a pixel in LR image $L$, and $\mathcal{P}$ be its surrounding patch (e.g., $3 \times 3$); then there exist similar patches $\mathcal{P}_1,...,\mathcal{P}_k$ (at sub-pixel shifts). These patches can be treated as if taken from $k$ different LR images of the same HR scene, thus inducing $k$ times more linear constraints on the values of pixels within neighborhood of $q \in H$. The solution of a system of $k$ linear constraints for each pixel results in the final estimation of high-resolution image $H$. For increased numerical stability, each equation induced by a patch $\mathcal{P}_i$ is globally scaled by the degree of similarity of $\mathcal{P}_i$ to its source patch $\mathcal{P}$. Thus, patches of higher similarity to $\mathcal{P}$ have a stronger influence on the recovered HR patch pixel values than patches of lower similarity.

## 2.1.2   EXAMPLED-BASED (ACROSS-SCALE) SUPER-RESOLUTION

The process outlined in Section 2.1.1 extends the application of classical SR to a single image. However, it suffers from all the limitations of classical SR (e.g., limited accuracy of patch registration, insufficient observations/patches, very low image resolution). These limitations of classical SR inspired the development of example-based super-resolution methods. Example-based SR methods learn correspondences between LR and HR image patch pairs from a database/dictionary and apply them to a new LR image to estimate its most likely high-resolution version. This same principle can be applied to a single image for super-resolution, but *without* using any external database. The LR-HR patch correspondences can be learned directly from the image itself by employing patch recurrence across multiple image resolution scales.

Let $B$ be the PSF relating the LR input image $L$ with the unknown HR image $H$: $L = (H * B) \downarrow s$. Let $I_0, I_1, ..., I_n$ denote a cascade of unknown images of increasing resolutions ranging from $L$ to the target HR image $H$ ($I_0 = L$ and $I_n = H$), with a corresponding cascade of PSFs $\{B_0, B_1, ..., B_n\}$ (where $B_n = B$ is the PSF relating $H$ to $L$, and $B_0$ is the $\delta$ function), such that every $I_l$ satisfies: $L = (I_l * B_l) \downarrow s_l$. Here $s_l$ denotes the relative scaling factor. Figure 2.3 illustrates the resulting cascade in purple. Often, the unknown PSF $B_l$ can be assumed to be Gaussian in nature, so that $B_l = B(l)$ form a cascade of PSFs with their variances determined by $s_l$. Let $L = I_0, I_{-1}, ..., I_{-m}$ denote a cascade of decreasing resolution obtained from $L$ using the same PSFs $B_l$: $I_{-l} = (L * B_l) \downarrow s_l$ ($l = 0,1,...,m$). Figure 2.3 illustrates the resulting cascade of known LR images in cyan color.

Let $\mathcal{P}_l(p)$ denote a patch in the image $I_l$ at pixel location $p$. The surrounding patch $\mathcal{P}_0(p)$ around any pixel $p \in L$ ($L = I_0$) in the input image can be searched for similar patches within the cascade of LR images $\{I_{-l}\}$, $l > 0$. Let $\mathcal{P}_{-l}(\tilde{p})$ be such a matching patch found in the LR image $I_{-l}$. Then its HR parent patch, $Q_0(s_l \cdot \tilde{p})$, can be extracted from the input image $I_0 = L$ (or from any intermediate resolution level between $I_{-l}$ and $L$, if desired). This provides an LR-HR patch pair $[\mathcal{P}, \mathcal{Q}]$, which provides a prior on the appearance of the HR parent of the LR input patch $\mathcal{P}_0(p)$, namely patch $Q_l(s_l \cdot \tilde{p})$ in the HR unknown image $I_l$. These steps can be expressed as (kindly refer to Figure 2.3 for an illustration): $P_0(p) \xrightarrow{findNN} P_{-l}(\tilde{p}) \xrightarrow{parent} Q_0(s_l \cdot \tilde{p}) \xrightarrow{copy} Q_l(s_l \cdot p)$. Here *NN* refers to the nearest neighbor of the patch under consideration.

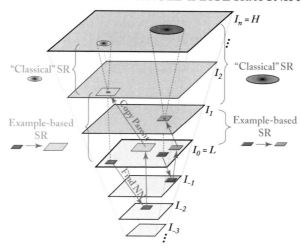

**Figure 2.3: Unified super-resolution framework combining Classical SR and Example-based SR**:
Patches in the input LR image $L$ are searched for similar patches in the down-scaled versions of $L$
(cyan images); when a similar patch is found, its parent patch is copied to the appropriate location
in the unknown HR image (purple images) with an appropriate gap in scale. A learned (copied) HR
patch induces classical SR linear constraints on the unknown HR intensities in the target image $H$.
The size of the blur kernels is determined by the residual scale gap between the resolution scale of
copied HR patches and resolution level of $H$. Image courtesy of Glasner et al. [42]

### 2.1.3    UNIFYING CLASSICAL AND EXAMPLE-BASED SR

The application of the process described in Section 2.1.2 for all the pixels in $L$ yields a large set
of (possibly overlapping) HR patches $\{Q_l\}$ between resolutions levels *(l = 0,1,...,n)*. Every such
learned HR patch $Q_l$ acts as an LR patch (blurred and subsampled) to its corresponding HR
patch in $H$, and thus induces linear constraints on unknown HR image $H$. These constraints
are of classical SR form described in Section 2.1.1, but induced by more compact blur kernel
$B_{n-l}$ than $B$ as they need to compensate only for the residual gap in scale *(n-l)* between the
resolution level $l$ of the learned patch and the final resolution level $n$ of the target HR image $H$
(see Figure 2.3).

The learned patches closer to the target resolution $H$ induce better conditioned constraints.
Each such linear constraint is globally scaled by its reliability (determined by its patch similarity
score) as described in Section 2.1.1. Note that if, for a particular pixel, its similar patches are found
only within $L$, then this scheme reduces to single-image classical SR described in Section 2.1.1;
if no similar patches are found, this scheme reduces to simple deblurring at that pixel. Thus,
the above scheme performs never worse than simple upscaling (interpolation) of $L$. The final
achievable magnification factor is determined by the depth of the LR image cascade.

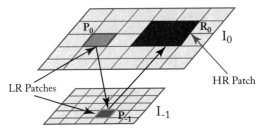

**Figure 2.4:** **Self-learning dictionary creation**: The LR-HR patch pairs are formed from the LR image and its coarser resolution version. There need not be a correspondence between the spatial positions of $P_0$ and $P_{-1}$.

## 2.2   SELF-LEARNING AND DEGRADATION ESTIMATION

The construction of the entire LR cascade and the use of interpolation in Glasner et al. [42] adds undesired complexity to an otherwise simple and elegant solution. It is possible to obtain an SR image with higher magnification factors with only an input LR image and its single coarser resolution version. The HR patches, usually interpolated in Glasner et al. [42], can be estimated using the already learned LR-HR patch pairs from the input LR image itself. This method of obtaining HR patches from the LR image itself is called *self-learning*. The readers should keep in mind that the only similarity shared by self-learning and the method explained in Section 2.1 [42] is their use in exploiting patch recurrence property of natural images. While Glasner et al. [42] use a unified framework of classical and example-based SR for HR reconstruction, self-learning is, at its core, an example-based SR, as it uses the sparsity of learned dictionary patch pairs to estimate HR patches.

Let $I_0$ be an input LR image and $I_{-1}$ be its coarser resolution version ($\downarrow 2$ for the current discussion), both captured using the same camera as illustrated in Figure 2.4. Let each block in the grid in $I_0$ and $I_{-1}$ represent a patch ($5 \times 5$ in the illustration). For any patch $P_0$ in LR image $I_0$, let $P_{-1}$ be a similar (or the same) patch found in $I_{-1}$. Let $R_0$ (of size $10 \times 10$) be a parent HR patch of $P_{-1}$ in $I_0$. Thus, $P_0$ and $R_0$ form an LR-HR patch pair, and $P_{-1}$ is the *best-mapped* patch for $P_0$. Repeating this procedure for a patch around all the pixels generates a set of LR-HR patch pairs. While searching for patches in $I_{-1}$, two patches are considered similar with the least value of *sum-of-squared-difference (SSD)*, below a certain threshold. Self-learning sets the SSD to *0* (exact match) to find the best-mapped patches. So, unlike in Glasner et al. [42], estimating the sub-pixel shifts is not required in self-learning.

The experimental observations indicate that only a small set of HR patches in $I_0$ act as parent patches for LR patches when looking for an exact match. Figure 2.5 presents relevant statistics for this observation. Experiments were conducted on 50 images with three different resolutions ($X, 2X$ and $4X$). Images with resolution of $128 \times 128$ and patch size of $5 \times 5$ have 15,376 patches when a patch around each pixel (boundary pixels excluded) is considered. These experiments in-

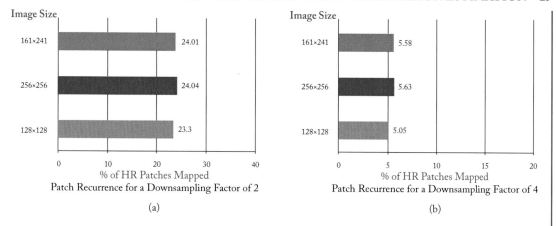

Figure 2.5: **Best-mapped patches:** Only a small set of LR patches find best-mapped patches within a coarser resolution vesion. These LR patches together with their corresponding HR patches form self-learning dictionaries. Image courtesy of Khatri and Joshi [62].

dicated that about 23% patches find best matches for a downsampling factor of 2; while the same number reduces to only about 5% for a downsampling factor of 4 as indicated in Figure 2.5. These *best-mapped* LR patches and their corresponding HR patches form the dictionaries. The HR patches corresponding to LR patches that could not find a match are estimated using these dictionaries. In the method proposed by Glasner et al. [42], which uses the *Approximate Nearest Neighbor (ANN)* algorithm [4] to find similar patches considering SSD as the measure for similarity, the remaining HR patches are obtained by interpolation of the LR patches. Their method subsequently uses these HR patches with the classical super-resolution approach to obtain the final super-resolved image. Since the super-resolution is performed using interpolated patches when a patch does not find any match, the final super-resolved image suffers in quality due to the aliasing effect.

In self-learning, however, the LR patches which do not find any match in the cascade are not simply interpolated; instead, two dictionaries (called the *self-learning* dictionaries), one comprising LR patches from $I_0$ with a match in $I_{-1}$, and the other comprising corresponding HR patches are constructed. The HR patches are available in the LR image $I_0$ itself, and are obtained using the best-mapped LR patches from its coarser resolution version $I_{-1}$. The HR patches corresponding to the remaining LR patches are estimated as follows: Let $N_b$ be the number of LR patches each of size $p \times p$ and each with the best match. Here the best match corresponds to a patch with minimum SSD. Let $p^2 \times 1$ denote a vector obtained after ordering the pixels of all $N_b$ patches lexicographically. So, the HR patch vector corresponding to $N_b$, for a magnification factor of $q$, would be of size $q^2 p^2 \times 1$. Then, the LR dictionary $D_l$ is of size $p^2 \times N_b$ and the corresponding HR dictionary $D_h$ is of size $q^2 p^2 \times N_b$, where $q = 2$ for the current discussion. It means there are $p^2$ and $q^2 p^2$ atoms (vectors) in LR and HR dictionaries, respectively.

Now, according to Compressive Sensing (CS), a theory in signal processing for sparse signal recovery [28], an LR patch from $I_0$ that does not have a match in $I_{-1}$ can be represented as a sparse linear combination of the patches from LR dictionary. This is based on the key assumption of CS theory that the most signals arising in nature are sparse, and can be represented as a linear combination of dictionary atoms using only a few non-zero coefficients. Suppose $\mathbf{x}$ is an unknown image vector in $\mathbb{R}^N$ with $N$ pixels. But, if we know *a priori* that $\mathbf{x}$ is compressible in a certain transform domain (e.g., wavelet, Fourier), i.e., if $\mathbf{x}$ is sparse, then as per CS theory, $\mathbf{x}$ can be acquired by measuring only $M$ linear projections rather than all $N$ pixels. If these projections are properly chosen, the size of $M$ (measurements) projections can be smaller than the size of image, $N$. For a $K$-*sparse* signal, these random projections work if $M = O(K \log(N/K))$. Under the sparsity assumption, a signal $\mathbf{x} \in \mathbb{R}^N$ can be recovered using $l_1$-minimization with the aid of standard optimization tools, such as linear programming, i.e.,

$$\min_{\mathbf{x} \in \mathbb{R}^N} ||\mathbf{x}||_{l_1}, \quad \text{subject to } \mathbf{y} = \phi\mathbf{x}; \quad \text{where } ||\mathbf{x}||_{l_1} = \sum_{i=1}^{N} |\mathbf{x}_i|. \tag{2.1}$$

Here $\mathbf{y}$ and $\phi$ represent the measured vector and the measurement matrix, respectively [11].

The coefficients for sparsely representing an LR patch from $I_0$ (with no match in $I_{-1}$) as a sparse linear combination of the patches from LR dictionary are estimated by formulating the problem as given in Equation 2.1. Dictionaries $D_l$ and $D_h$ are used as projections in order to obtain the unknown HR patches. Here $\mathbf{y}$ corresponds to an LR patch, $\mathbf{x}$ is an unknown sparse co-efficients vector and $D_l$ represents the measurement matrix $\phi$. Assuming the sparsity to be the same for LR and HR patches, the estimated sparse coefficients are used to obtain the corresponding HR patch as a sparse linear combination of HR patches of $D_h$ as illustrated in Figure 2.6. Repeating the above procedure for all the LR patches that do not find a match gives us the complete set of LR-HR patch pairs. However, the HR patches obtained so far still lack contextual dependencies. Nevertheless, these LR-HR patch pairs are used to estimate the degradations. Let $\mathbf{y}_i$ and $\hat{\mathbf{z}}_i$ represent the $i^{th}$ LR and its corresponding HR patch in the LR image ($I_0$), ordered lexicographically. Let $B_i$ be the unknown degradation for the $i^{th}$ patch. Then, the LR patch can be modeled as:

$$\mathbf{y}_i = B_i \hat{\mathbf{z}}_i + \mathbf{n}_i \tag{2.2}$$

where $\mathbf{n}_i$ represents the *independent and identically distributed (i.i.d.)* Gaussian noise vector with variance $\sigma_n^2$. Note that $B_i$ is essentially a blur and downsampling matrix. For every LR-HR patch pair, the corresponding degradation $B_i$ is estimated using a simple non-negatively constrained least squares approach [18]. These estimates are then used in a regularization framework to obtain the final super-resolved image.

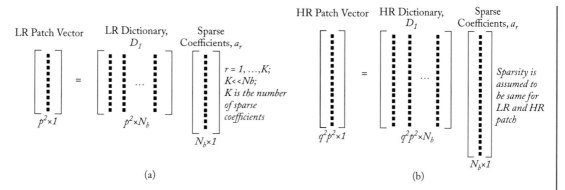

Figure 2.6: **Illustration of the use of CS theory for recovery of missing HR patches using self-learning dictionary $D_h$ and LR dictionary $D_l$:** sparse coefficients obtained from (a) are used in (b) to obtain the unknown HR patch. Image courtesy of Khatri and Joshi [62].

## 2.3 GABOR PRIOR AND REGULARIZATION

Edge-preserving priors tend to recover edges sufficiently well to enhance the perceptual quality of the super-resolved image. They are suitable as long as the details are present as high frequency contents. However, an image may have various textured regions with different frequencies, and hence, all these details should be preserved in the super-resolved image. This can be achieved with a prior that incorporates the information about the details at various frequencies during the reconstruction process of the super-resolved image.

In computer vision, the Gabor filter is a linear filter used for feature extraction at different frequencies and orientations. The frequency and orientation representations of Gabor filters are similar to those of certain neurons in the Human Visual System, and they have been found to be particularly appropriate for texture representation and discrimination. In the spatial domain, a 2D Gabor filter represents a Gaussian kernel function modulating a 2D sinusoid. Hence, the *PSF* of the filter is given by [99],

$$g\left(x, y, f, \theta, \sigma_x, \sigma_y\right) = e^{-\frac{1}{2}\left(\frac{x'^2}{\sigma_x^2} + \frac{y'^2}{\sigma_y^2}\right)} \cos(2\pi f x') \qquad (2.3)$$

where $(x, y)$ represents the spatial coordinates; $(x', y') = (x \cos\theta + y \sin\theta, -x \sin\theta + y \cos\theta)$ denotes the rotated $(x, y)$ coordinates; $\sigma_x$ and $\sigma_y$ denote the spatial extent of the filter in $x-$ and $y-$ directions, respectively; $f$ is the frequency of sinusoid and $\theta$ is its orientation. Equation 2.3 is a mathematical representation of a $2D$ Gaussian bandpass filter. The Gabor prior based on the outputs of Gabor filters can be explained as follows: a Gabor filter bank is constructed for extracting features at various frequencies and orientations from a given image. These filters are applied over (a) the input LR image and (b) the simulated LR image obtained by blurring and down-sampling the current SR estimate using estimated degradation $B_i$. The prior imposes a condition

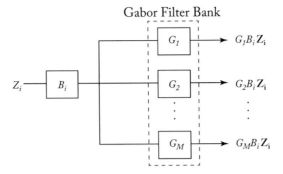

**Figure 2.7: Gabor prior for HR patch** $Z_i$: $G_j B_i Z_i$ produces an image feature at a particular frequency band which has to match the feature of $y_i$ when it is passed through the same filter $G_j$; $G_j$, $j = 1,..., M$ represents a Gabor filter bank. Image courtesy of Khatri and Joshi [62].

for similarity of features between the simulated LR image and LR input image when viewed at different frequencies. This means the downsampled version of an HR estimate should have the same Gabor features to that of the LR input image when passed through a Gabor filter bank, as illustrated in Figure 2.7.

The HR patches estimated in the aforementioned manner fail to provide the necessary spatial dependencies between high-resolution pixels due to patch-based estimation. Therefore, it is necessary to regularize the solution in order to take care of the spatial correlation among the neighboring pixels in the final super-resolved image. The regularization approach requires a data-fitting term, and a prior term that seeks to improve the solution based on the *a priori* knowledge about it. A vast variety of priors have been attempted whenever the problem is formulated in a regularization framework. Most of these priors are based on Markovian property, imposing spatial dependencies between the pixels in the final solution [47, 117]. Using a data-fitting term and a regularization term comprising the Gabor prior, the final cost function to be minimized can be written as,

$$\epsilon = \sum_{i=1}^{N} \left( ||\mathbf{y}_i - B_i \mathbf{z}_i||^2 + \sum_{j=1}^{M} ||G_j \mathbf{y}_i - G_j (B_i \mathbf{z}_i)||^2 \right). \tag{2.4}$$

Here $\mathbf{y}_i$ is the $i^{th}$ LR patch, $B_i$ represents the degradation matrix estimated using the $i^{th}$ LR-HR patch pair and $N$ is the total number of patches in LR image. $G_j$ represents the $j^{th}$ Gabor filter with the impulse response represented as in Equation 2.3, $M$ represents the total number of filters in the filter bank and $\mathbf{z}_i$ is a high-resolution patch corresponding to LR patch $\mathbf{y}_i$. This cost function is convex and can be easily minimized using the gradient descent optimization method. In order to speed up the convergence, the $\hat{\mathbf{z}}_i$ ($i = 1, 2, ..., N$) estimated with self-learning is used as the initial estimate. Minimization of the above cost function yields the final super-resolved image $\mathbf{z}$. Since the LR-HR pairs have a resolution difference of 2, the estimated SR image is twice the size of the LR image. In order to obtain the SR for a magnification factor of 4 or more, the estimated SR is considered as the LR image, and the entire process of learning and regularization is repeated. As an alternative, one can use an input image and its coarser resolution version with larger resolution difference (> 2) to achieve higher magnification factors.

## 2.4   PERFORMANCE EVALUATION

In this section, the performance of self-learning and Gabor prior is evaluated both perceptually and quantitatively. However, the assumptions and environment within which these results are produced have to be defined at the outset. The SSD difference for a patch to be best-mapped is set to *0*. The input image is assumeed to be "clean," i.e., the input image should have a high signal-to-noise ratio. As the dictionary learning relies heavily on the exact match for patches, the algorithm may produce unsatisfactory results for a noisy input image or may even fail completely. So, necessary pre-processing should be performed before super-resolving the image. All the results presented are obtained by working on grayscale images. For color images, RGB components are transformed to $YC_bC_r$, and only the luminance ($Y$) component is super-resolved. $C_b$ and $C_r$ components are upsampled using bicubic interpolation. For Gabor filtering, a filter bank comprising eight filters is chosen with parameters listed in Table 2.1. These chosen filter parameters are for the results presented in this section. Different combinations of frequencies and orientations were considered; however, the chosen parameters seemed to provide a satisfactory balance between the execution time and the output image quality. A set of parameters that produced perceptually and quantitatively consistent results for a variety of images is chosen. The optimization is performed using the gradient descent method. The step size, $\lambda$, for gradient descent optimization is chosen to be 0.1. The $l_1$-*MAGIC* solver [12] is used to perform $l_1$-minimization. Bicubic interpolation and Glasner et al. [42] are used for comparison.

Table 2.1:  Gabor filter parameters

| Frequencies | Orientations (in degrees) |
|---|---|
| 0.05, 1.75 | 0, 45, 95, 135 |

### 2.4.1   QUALITATIVE EVALUATION

Heritage monuments often have intricate details with very fine grainy texture over smooth areas. So if a detail-preserving prior oversharpens such fine texture, it can make the image look aesthetically unpleasant. Figure 2.8 shows the super-resolved results for the images of a heritage site in Hampi, India, for the magnification factor of 2. The LR images displayed in Figure 2.8a have varying levels of textural details. Figure 2.8d exhibits the effectiveness of self-learning by preserving the intricate carvings on the monument walls. Figures 2.9–2.11 present the super-resolved images for higher magnification factors. For relatively smoother images, the difference between

(a) LR      (b) Bicubic Interpolation      (c) Glasner et al.      (d) Self-learning + Gabor

**Figure 2.8: Super-resolution for 2X [Carving (375x500) and Ramachandra (400x300)]:** (a) LR; (b) Bicubic interpolation; (c) Glasner et al. [42]; (d) Self-learning + Gabor Prior. The top row has smooth texture except the carvings. Both Self-learning and Glasner et al. [42] produce visually similar results in this case. However, notice the darker regions between the elbows and thighs of the carving which is much more emphasized in (c) and (d). A similar effect can be perceived in the bottom row between the transition from the darker to the brighter regions.

the results of Glasner et al. [42] and self-learning could be very similar (see Figure 2.11). However, for images with detailed textures or complex details, the power of self-learning comes to the fore. The bottom row of Figure 2.9 is an extremely complex scene of the Taj Mahal interior with an interplay between light and shadows. A closer inspection of the light spots on the wall in Figures 2.9c and 2.9d reveal the brightness (and, in effect, the sharpness) of the spots in the latter. This would help the reader understand and appreciate the simple, yet effective, way in which self-learning in conjunction with the Gabor prior achieves its objective.

## 2.4.2 QUANTITATIVE EVALUATION

Often the reference image would be unavailable for quantitative evaluation of a method. So evaluation measures like power signal-to-noise ratio (PSNR), structural similarity (SSIM) index [128] or feature-similarity (FSIM) index [138] that require a reference image cannot be applied to eval-

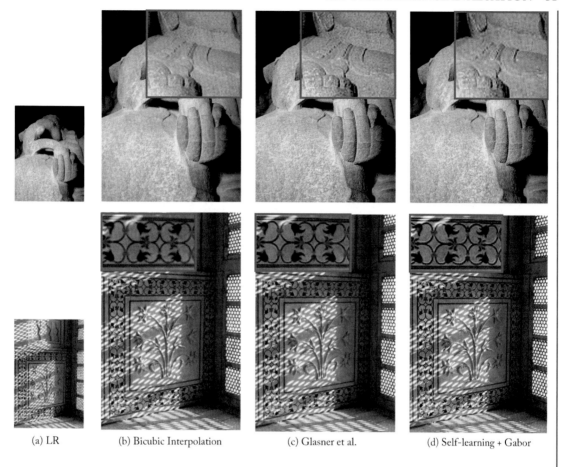

(a) LR                (b) Bicubic Interpolation          (c) Glasner et al.          (d) Self-learning + Gabor

Figure 2.9: **Super-resolution for 4X [Ganesh** (432x324**) and Taj** (480x318**)]:** (a) LR images;(b) Bicubic Interpolation; (c) Glasner et al. [42]; (d) Self-learning + Gabor Prior.

uate the results presented in this section. So Blind/Referenceless Image Spatial QUality Evaluator (BRISQUE) [82] is used for measuring the image quality. Table 2.2 lists the BRISQUE scores for the results presented in this section. An interesting entry in the table belongs to the score for $8X$ (Figure 2.11). Though self-learning and Glasner et al. [42] look perceptually much sharper, their respective BRISQUE scores are nearly half that of bicubic interpolation. There are possibly two reasons for this: one, due to extremely low resolution of the input image, the self-learning dictionary does not have enough high detail patches to reconstruct the intricate carvings that form the major part of the scene structure; two, BRISQUE looks for the spatial consistency of image pixels, which is high for bicubic image due to over-smoothing.

Although conceptually the self-learning does show promise to be an effective super-resolution technique, it should also be scrutinized for its operational limitations. It requires the creation of dictionaries of matched patches. For high spatial resolution images, these dictionaries could take up a large amount of memory, which may not be on offer every time. Also, $l_1$-minimization is computationally a very intensive process which would further diminish the performance. A redressal of these issues would be necessary for self-learning to be considered more than just an impressive concept. Chapter 3 attempts to mitigate these issues to increase the utility of self-learning.

Table 2.2: BRISQUE scores (Higher score indicates better image quality.)

| | 2X | | 4X | | | | 8X | |
| | Carving | Ramachandra | Hampi | Modhera | Ganesh | Taj | Sidi | Average Score |
|---|---|---|---|---|---|---|---|---|
| Bicubic | 39.19 | 50.06 | 55.36 | 56.05 | 56.62 | 54.67 | **65.15** | 54.03 |
| Glasner et al. [42] | **57.15** | 60.26 | 55.32 | 54.27 | 64.29 | **57.21** | 36.29 | 54.97 |
| Self-Learning | 53.99 | **62.21** | **56.78** | **58.17** | **71.30** | 55.12 | 35.14 | **56.10** |

## 2.5   CONCLUSION

Self-similarity is an important characteristic of the artistic work found in monuments at heritage sites. Self-learning exploits this self-similarity of patches to obtain LR-HR patch pairs from the given image itself. The Gabor prior that attempts to emulate the details discrimination behavior of the Human Visual System regularizes the estimated super-resolved image to be visually pleasant.

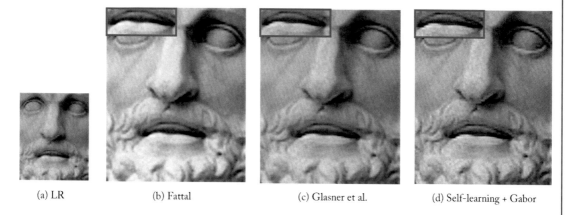

(a) LR          (b) Fattal          (c) Glasner et al.          (d) Self-learning + Gabor

**Figure 2.10:   Super-resolution for 8X [Sculpture (85x70))]:** (a) LR image; (b) Fattal [32]; (c) Glasner et al. [42]; (d) Self-learning + Gabor Prior. Notice the color tone of the result of Fattal [32] and Self-learning. Self-learning keeps the color tone intact. The BRISQUE scores for (b)–(d) are 72.17, 61.50 and 69.66, respectively. Image courtesy of Khatri and Joshi [62].

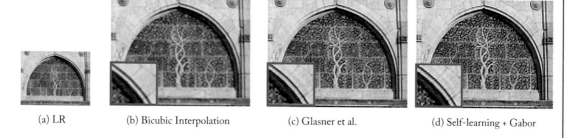

(a) LR          (b) Bicubic Interpolation          (c) Glasner et al.          (d) Self-learning + Gabor

**Figure 2.11:   Super-resolution for 8X [Sidi (101x135)]:**(a) LR; (b) Bicubic Interpolation; (c) Glasner et al. [42]; (d) Self-learning + Gabor Prior.

CHAPTER 3

# Self-learning: Faster, Smarter, Simpler

Self-learning, introduced in the previous chapter, uses only one image to achieve super-resolution with high magnification factors. But, as the image resolution increases, the number of patches in the dictionary also increases dramatically, and makes the SR reconstruction computationally prohibitive. It employs $l_1$-minimization to exploit the sparsity inherent within natural images, which further compounds the time-complexity problems associated with the technique. One way to mitigate this problem is to reduce the dictionary size in a meaningful way by employing some kind of learning algorithms (e.g., K-SVD, K-means). However, such learning algorithms come with their own computational burden, and may ultimately prove to be of little advantage when time is of the essence. Another way to find the best-mapped patch for a particular LR patch is to restrict the search space within a neighborhood of the LR patch in the coarser-resolution image. But this may not always result in the best-mapped patch for that particular LR patch.

This chapter introduces an improvement over the method discussed in Chapter 2 and takes the approach that falls in the middle of the two approaches suggested above. It removes the redundancies inherent in large self-learned dictionaries to super-resolve an image. It neither uses any computationally taxing learning methods nor constrains the search space for LR patches, and yet manages to super-resolve an image much more efficiently compared to the method suggested in Chapter 2. The method uses patch variance to reduce the dictionary size and speed up the super-resolution process. It also does away with the Gabor prior or other forms of regularization to simplify the estimation process. Later in the chapter, it will be established that any low-variance (fewer details) patch that does not find any match can be represented as a linear combination of only low-variance patches from the dictionary, i.e., low-variance LR patches from the dictionary suffice to represent any non-dictionary LR patch. The same principle applies to high-variance (high details) patches. Images with high magnification factors can be obtained with this method without *any regularization or prior information*, which can be subjected to further regularization with necessary prior(s) to refine the super-resolved image.

## 3.1 EFFICIENT SELF-LEARNING

The method discussed in this chapter, because of its simplicity and improvements suggested over self-learning with sparsity, may prove to be a useful tool for quick super-resolution of images captured on mobile phones. Unlike self-learning explained in Chapter 2, that builds large-sized

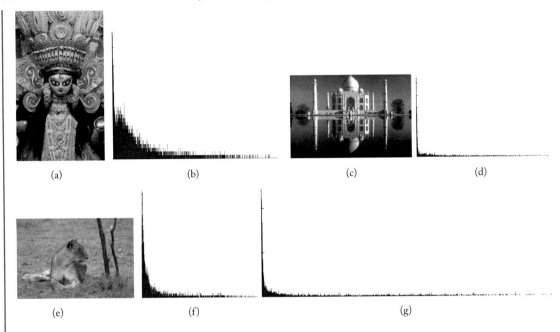

Figure 3.1: **Variance histogram of LR dictionary patches**: (a) idol; (c) heritage; (e) natural scene; (b)–(d)–(f) histograms of variances; (g) combined variance histogram of all 1,300 images. The majority of the image dictionaries have a long-tailed variance distribution. The concentration of histograms near the origin indicates a large number of low-variance patches. In the plots, x- and y-axes indicate variances and their frequency of occurence, respectively.

LR-HR dictionaries (for a reasonably large image, thereby increasing the time complexity), this approach reduces the dictionary size by taking only a small percentage of the total best-mapped dictionary patches, and further segments it into three smaller dictionaries, namely, low-variance, high-variance, and medium-variance dictionaries. The non-dictionary patches (i.e., image patches that do not find any best match) are obtained from these reduced dictionaries. The method has two core components: 1) dictionary creation and 2) dictionary redundancy removal.

### 3.1.1    IMPROVED SELF-LEARNING FOR SUPER-RESOLUTION

Self-learning discussed in Chapter 2 considers exact matching patches as the candidates for LR-HR dictionaries; though, for large images, these dictionaries become extremely large. The experimental observation of variance distribution (histogram of variances) of LR dictionaries for around 1,300 images (including paintings, abstract art, man-made structures and natural images) indicates that the majority of LR dictionary patches have very low-variance. Figure 3.1 shows some of the sample images along with the variance histogram of their corresponding LR dictionary

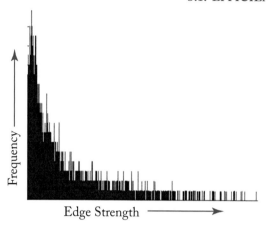

**Figure 3.2: Histogram of edge strengths of LR dictionary patches**: The shape of edge-strengths (gradient magnitude) distribution resembles that of variance distributions of dictionary patches in Figure 3.1. This observation allows reconstruction of a patch of a particular variance class as a linear combination of patches from the same variance class. Image courtesy of Khatri and Joshi [63].

patches, particularly Figure 3.1g, which shows the combined histogram of all 1,300 images consisting of images downloaded from the Internet and Martin et al. [79]. This empirical evidence provides a key insight about the uniformity of details-distribution in different types of images and can be used to create LR-HR dictionaries in a more efficient way.

**Dictionary Creation**

Self-learning employs $l_1$-minimization to solve for non-dictionary patches that not only have very high time complexity, but the LR dictionary itself has inherent redundancy of the patches. These redundancies can be eliminated in a simple and efficient manner by employing much faster $l_2$-minimization instead of using computationally taxing $l_1$-minimization. In the experiments carried out, it has been observed that, based on the variance distribution of LR dictionary patches, any low-variance non-dictionary patch can be reconstructed by a linear combination of only low-variance dictionary patches. Similarly, for high- or medium-variance non-dictionary patches, a linear combination of patches with high- or medium-variances would suffice. This observation can be supported by considering low-variance patches as patches lacking in details and high-variance patches as patches containing high-details (edges). For this assumption to be valid, the histogram of edge-strengths (gradient magnitude for each dictionary patch) of LR dictionary patches should resemble that of the histograms in Figure 3.1. Figure 3.2 shows the histogram of edge-strengths of LR dictionary patches for one of the sample images. It validates the assumption that high-variance patches contain details and vice versa. This observation forms the basis for the variant of

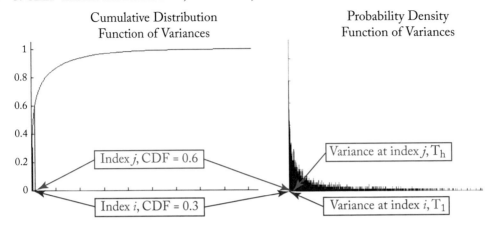

**Figure 3.3: Selection of thresholds for dictionary division**: A sharp rise in the CDF curve near the origin indicates a very high density of low-variance patches. Indices $i$ and $j$ in CDF are used to select $T_l$ and $T_h$ from PDF of variances. $T_l$ and $T_h$ are variables. Image courtesy of Khatri and Joshi [63].

self-learning introduced in Chapter 2. Here the self-learned LR dictionary is divided into three separate dictionaries comprising low-, medium-, and high-variance patches, separately.

The *Cumulative Distribution Function* (CDF) of LR dictionary variances (Figure 3.3) provides a guideline for the division of the LR dictionary. It represents the process of threshold selection for the LR dictionary division. The variance value in the *Probability Density Function* (PDF) of variances plot that has a cumulative value of 0.3 with all the preceding variance values is chosen as low-variance threshold, $T_l$ (index $i$ in Figure 3.3). Similarly, the variance value in the PDF that has a cumulative value of 0.6 with all the preceding variance values is chosen as high-variance threshold, $T_h$ (index $j$ in Figure 3.3). Any LR dictionary patch with variance less than $T_l$ is classified as a low-variance patch, and any LR dictionary patch with variance greater than $T_h$ is classified as a high-variance patch; while the rest of the dictionary patches are termed as medium-variance patches. The typical values for $T_l$ and $T_h$ are *0.3* and *0.6*, respectively. It is to be noted that $T_l$ and $T_h$ values have been chosen empirically after trying different values with only marginal improvements in super-resolved images, and the chosen $T_l$ and $T_h$ values provide an acceptable division of LR dictionary patches. The threshold selection process can be automated; but this may lead to a compromise in the speed of the method without gaining much in terms of super-resolved image quality.

### Dictionary Redundancy Removal
Even after the division, the low-variance dictionary still exhibits redundancies among its patches. These redundancies can be eliminated by a simple *k-nearest neighbor* ($k = 1$) operation. So, for every patch in the low-variance dictionary (without a stringent constraint of finding the exact

2,995 Patches 2,119 Patches

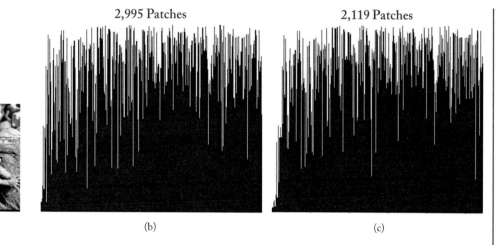

(a) (b) (c)

**Figure 3.4: Redundancy elimination for low-variance dictionary**: (a) Input image; (b) variance distribution *before* redundancy removal; (c) variance distribution *after* redundancy removal. The shape of the distribution is preserved pre- and post-redundancy removal with 30% reduction in low-variance dictionary patches. Image courtesy of Khatri and Joshi [63].

match), its most similar patch is searched within the low-variance dictionary itself. For every such LR patch pair found, the LR patch with a higher variance value is retained in the dictionary, while the other LR patch of the pair is discarded from the dictionary. Since medium- and high-variance represent details of the image, this operation is performed only on the low-variance dictionary. Figure 3.4 shows a dictionary variance distribution *before* and *after* redundancy removal. The HR dictionary is formed by extracting for every LR patch its corresponding HR patch from the HR dictionary. Thus, three LR dictionaries and their corresponding HR dictionaries are constructed. After obtaining LR dictionaries in the aforementioned manner, non-dictionary LR patches are represented as a linear combination of patches from one of the three LR dictionaries depending upon their variances, and the corresponding coefficients for each non-dictionary LR patch are obtained using a *least-norm* minimization, as in Equation 3.1, where $\mathbf{y}$ is an LR patch vector; $L_{lv}$, $L_{mv}$, $L_{hv}$ are low-variance, medium-variance and high-variance LR dictionaries, respectively; $\mathbf{x}$ is a coefficient vector. Equation 3.1 represents an *under-determined* system of equations. Hence, there is no unique solution in the least-squares sense.

$$min.\|\mathbf{x}\|_2; \text{ s.t. } \mathbf{y} = \begin{cases} L_{lv}\mathbf{x} & \text{if } Var(\mathbf{y}) \leq T_l \\ L_{mv}\mathbf{x} & \text{if } T_l < Var(\mathbf{y}) \leq T_h \\ L_{hv}\mathbf{x} & \text{if } Var(\mathbf{y}) > T_h \end{cases} \quad (3.1)$$

For every LR patch, its corresponding HR patch (based on its variance) can be reconstructed from one of the three HR dictionaries and the estimated coefficient vector $\mathbf{x}$, as given in Equation 3.2.

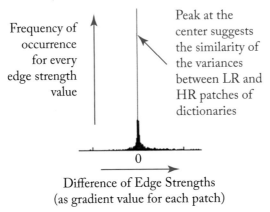

Frequency of occurrence for every edge strength value

Peak at the center suggests the similarity of the variances between LR and HR patches of dictionaries

0

Difference of Edge Strengths
(as gradient value for each patch)

**Figure 3.5: Distribution of difference of variances between LR-HR patches of dictionary**: Gaussian shape of the distribution along with a long tail indicates the similarity of variances between LR and their corresponding HR patches. Image courtesy of Khatri and Joshi [63].

Here $\mathbf{z}$ is a reconstructed HR patch; $H$ indicates HR dictionary and the subscript represents the dictionary class.

$$\mathbf{z} = \begin{cases} H_{lv}\mathbf{x} & \text{if } Var(\mathbf{y}) \leq T_l \\ H_{mv}\mathbf{x} & \text{if } T_l < Var(\mathbf{y}) \leq T_h \\ H_{hv}\mathbf{x} & \text{if } Var(\mathbf{y}) > T_h \end{cases} \tag{3.2}$$

A natural question at this stage could be the reason behind the use of co-efficients recovered using LR patches to estimate HR patches. Empirical evidence again suggests the appropriateness of the assumption. Figure 3.5 shows the histogram plot of the difference of variances between LR and their corresponding HR patches. It validates the use of estimated coefficients from LR patches to reconstruct corresponding HR patches. A spike at the origin hints at the "sameness" of low-variance and medium-variance patches, while the long tail of the distribution points at the similarity between the high-variance LR and HR patches. Most of the image dictionaries exhibit such behavior. The division of a dictionary into smaller dictionaries, along with the use of least-norm without any regularizing priors, makes this variant of self-learning simpler, smarter and computationally efficient in comparison to self-learning with sparsity. It is simpler and smarter in the sense that the division of patches across different dictionaries based on their variances leads to the requirement of fewer patches for reconstruction, while using least-norm greatly simplifies the non-dictionary patch reconstruction process. It is to be noted that the procedure explained above super-resolves the image by a factor of 2. Higher magnification factors can be achieved by considering the current super-resolved image as an LR image and repeating the above procedure.

(a) LR                    (b) Shane et al.           (c) Self-learning by Sparsity           (d) Self-learning by
                                                                                              Dictionary Reduction

**Figure 3.6:** **Upsampling by 2X [Castle (481x321) and Hampi (479x720)]:** (a) LR image; (b) Shan et al. [109]; (c) Self-learning by Sparsity; (d) Self-learning by Dictionary Reduction.

## 3.2    PERFORMANCE EVALUATION

The experimental setup for the performance evaluation of the ideas introduced in the previous section remain the same as that outlined for Chapter 2. Here $I_{-1}$ is obtained from $I_0$ by a blur and downsample process. Patches of size 3x3 are used. For self-learning, a nearest neighbor search is performed using *Approximate Nearest Neighbor* [4]. The threshold values, $T_l$ and $T_h$, for dictionary division are empirically set to *0.3* and *0.6*, respectively. An automatic way could be devised for the thresholds selection; but, due to the similarity of variance distribution across different kinds of images, using fixed threshold values saves precious computational time without severely affecting the image quality. Both the variants of self-learning (i.e., with sparsity, discussed in Chapter 2 and dictionary division) are implemented in MATLAB®.

### 3.2.1    PERCEPTUAL AND QUANTITATIVE EVALUATION

As already suggested in an earlier section, instead of being a complete super-resolution solution, self-learning could also be incorporated as a sub-entity in a larger super-resolution framework.

| (a) LR | (b) Shane et al. | (c) Self-learning by Sparsity | (d) Self-learning by Dictionary Reduction |

Figure 3.7: **Upsampling by 4X [Sidi (336x448) and Taj (480x318)]:** (a) LR Image; (b) Shan et al. [109];(c) Self-learning by Sparsity; (d) Self-learning by Dictionary Reduction.

This section evaluates the performance of self-learning without applying any regularization/edge-preserving priors. To achieve the objective of computational efficiency, only a portion of the total best-mapped dictionary patches is selected to estimate the final super-resolved image. These selected patches constitute the dictionary, and the division and reduction process is applied on them. The results are compared with Shan et al. [109] and self-learning with Gabor prior from Chapter 2. Figures 3.6–3.8 show the results for magnification factors of 2 to 8. Table 3.1 lists the BRISQUE scores for the same. Ideally, self-learning with Gabor prior should always outperform the results of raw self-learning with only a small percentage of the patches selected. However, interestingly, as shown in Table 3.1, on a couple of occasions, with images having a fairly repetitive texture, raw self-learning may even outperform regularized self-learning. Table 3.2 shows the dictionary statistics for the results in Figures 3.6–3.8. For most of the images, the results of self-learning with dictionary division would be comparable to that of sparsity. The last entry in Table 3.1 indicates that on average dictionary division not only outperforms sparsity, but produces

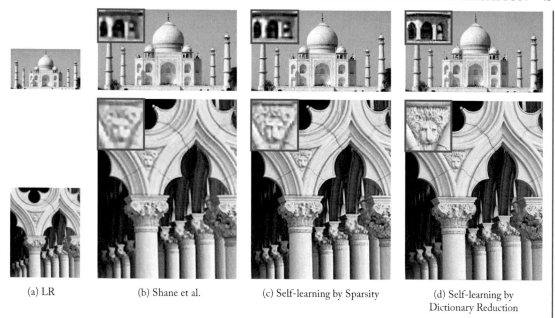

(a) LR            (b) Shane et al.            (c) Self-learning by Sparsity            (d) Self-learning by
                                                                                    Dictionary Reduction

**Figure 3.8: Upsampling by 8X [Tajmahal (127x225) and Venice (159x127)]:** (a) LR Image; (b) Shan et al. [109]; (c) Self-learning by Sparsity;(d) Self-learning by Dictionary Reduction. Self-learning with sparsity has a higher quality score, yet, self-learning in its simplified form has a much better visual appearance. "Venice" image courtesy of Kandice Dickinson.

a performance that matches that of Shan et al. [109]; Table 3.2 suggests that only a small portion of the total dictionary patches were sufficient to match the performance of Shan et al. [109].

There are two factors which ultimately dictate the performance of self-learning: one, the spatial resolution of the LR image, as the size of the final reconstruction dictionary depends on it; two, for images with sufficient spatial resolution, the percentage of patches selected would again decide the final SR quality. When the aforementioned criteria are satisfied, threshold values chosen for dictionary division do not have a severe effect on the results, except for a slight blurring at the edges when $T_l$ is high. Figure 3.9 demonstrates the case when the performance of self-learning would not be acceptable. Figures 3.9b–d show the effect of final reconstruction dictionary size and threshold values ($T_l$ and $T_h$) when spatial resolution is sufficient. Figure 3.9b shows artifacts near transition for the normal value of $T_l$. However, as Figures 3.9c–d indicate, the artifacts disappear as soon as the dictionary division is unequal between variances (notice the threshold values below the images). There is also a marginal improvement in BRISQUE scores when the proper threshold values are chosen. Figures 3.9f–g show the failure of self-learning even at its full capacity when the image has extremely low spatial resolution.

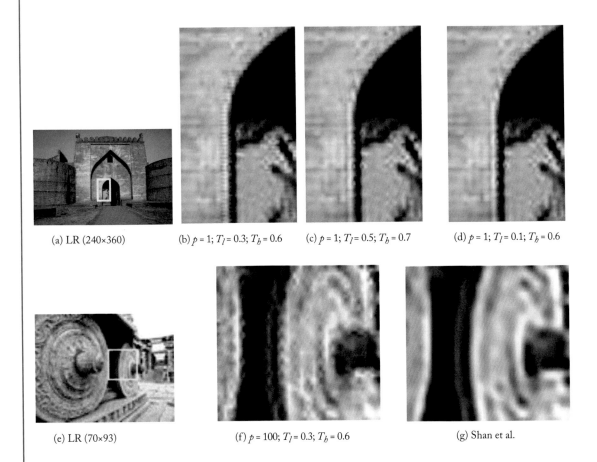

(a) LR (240×360)   (b) $p = 1$; $T_l = 0.3$; $T_h = 0.6$   (c) $p = 1$; $T_l = 0.5$; $T_h = 0.7$   (d) $p = 1$; $T_l = 0.1$; $T_h = 0.6$

(e) LR (70×93)   (f) $p = 100$; $T_l = 0.3$; $T_h = 0.6$   (g) Shan et al.

**Figure 3.9: Limitations of self-learning:** BRISQUE scores for (*top row*) (b) 37.03, (c) 37.13, (d) 37.68. $p$ is the percentage of best-mapped patches selected for reconstruction. Change in $T_l$ and $T_h$ removes artifacts but provides only marginal improvement in image quality. BRISQUE scores for (*bottom row*) (f) 41.41, (g) 56.87. Self-learning demonstrates pixelation effects for images with extremely low spatial resolution. Image in (a) has total of 16,180 best-mapped patches; (e) has total of 1,386 best-mapped patches.

Table 3.1: BRISQUE scores (Higher score indicates better image quality.) Self-learning in its reduced and simplified form has a performance comparable to other two super-resolution methods.

|  | 2X | | 4X | | 8X | | |
|---|---|---|---|---|---|---|---|
|  | Castle | Hampi | Sidi | Taj | Tajmahal | Venice | Average Score |
| Shan et al. [109] | 51.35 | **61.80** | **55.04** | 57.40 | 81.28 | 63.69 | **61.76** |
| Self-learning (Sparsity) | 43.06 | 43.63 | 47.28 | 55.12 | **84.36** | **73.60** | 57.84 |
| Self-learning (Dictionary Reduction) | **51.99** | 47.81 | 54.80 | **59.03** | 70.46 | 70.69 | 59.13 |

Table 3.2: Dictionary reduction statistics for Figures 3.6–3.8

|  | 2X | | 4X | | 8X | |
|---|---|---|---|---|---|---|
|  | Castle | Hampi | Sidi | Taj | Tajmahal | Venice |
| Original Dictionary | 30,680 | 77,092 | 4,31,994 | 1,17,332 | 55,393 | 55,297 |
| Percentage ($p$) Selected | 20.0 | 10.0 | 3.0 | 10.0 | 30.0 | 30.0 |
| Raw Dictionary | 6,136 | 7,709 | 12.960 | 11,733 | 16,618 | 19,889 |
| Low-variance Dictionary (*before reduction*) | 1,819 | 2,309 | 3,857 | 3,486 | 4,973 | 5,892 |
| Low-variance Dictionary (*after reduction*) (A) | 1,331 | 1,629 | 2,673 | 2,445 | 3,578 | 4,169 |
| Medium-variance Dictionary (B) | 1,846 | 2,307 | 3,907 | 3,529 | 4,965 | 6,008 |
| High-variance Dictionary (C) | 2,463 | 3,089 | 5,191 | 4,709 | 6,664 | 7,963 |
| Final Reconstruction Dictionary (A+B+C) | 5,640 | 7,025 | 11,771 | 10,683 | 15,207 | 18,140 |
| Total Reduction for A (%) | 8.00 | 8.87 | 9.17 | 8.95 | 8.49 | 8.79 |
| Effective Dictionary Size Selected | 18.38 | 9.11 | 2.73 | 9.11 | 27.45 | 27.36 |

## 3.2.2   IMPROVEMENTS AND EXTENSIONS

The creation of best-mapped patches' dictionary forms the back-bone that drives the performance of self-learning. The current version of self-learning works exclusively with pixel values of patches. A better criteria for the search of best-mapped patches could reduce the rate of false matches. One way to do this would be to use higher-level features along with pixel values while performing the search. The conversion to gray-scale can be skipped for color images, and color features/color values can be used for matching. Although dictionary division using variance values produces acceptable results, variance values are still highly sensitive to pixel values. Even a single noise pixel in an otherwise smooth patch would affect the final classification of that patch. Compression artifacts can also affect the performance. When a certain percentage of patches are selected for reconstruction, they are selected in the order they occur. It is possible that more suitable candidates for inclusion in the reconstruction dictionary are omitted as a result of this. A more scientific way of selecting the required patches can greatly enhance the performance of self-learning. So, depending upon the nature of the problem, various pitfalls mentioned above can have a different effect on the final outcome.

## 3.3   CONCLUSION

This chapter presented an interesting property-of-variance distribution of sufficiently small patches that is common across different kinds of images, and incorporated that into a self-learning framework. It also established that, with very few constraints, self-learning can be used for a quick super-resolution of images, and makes it a simple, yet effective, method that can be incorporated into a larger framework. The ideas presented here are used in subsequent chapters to solve the problem of image inpainting, which is an important part of digital heritage reconstruction and preservation.

CHAPTER 4

# An Exemplar-based Inpainting Using an Autoregressive Model

Digitized 3D models of heritage scenes/objects can be generated with the help of data captured using laser scanners or a large number of photographs. Due to self-occlusion or difficulty in capturing the scene/object from a particular viewpoint, some parts of the scene/object may not be captured at all. It may also happen that the monument is ruined or a part of it is damaged, leaving behind cracks. Such situations result in the creation of holes, i.e., missing information in the estimated 3D surface from the captured input data. Although Poisson surface reconstruction [60] can be used to fill the holes, it performs over-smoothing. If the surface is reconstructed from a large number of photographs, one can overcome this problem by performing resolution enhancement prior to estimating the 3D surface using the techniques discussed in Chapters 2–3. In this chapter, we present an inpainting method (i.e., a process of restoring or modifying the image contents imperceptibly) to fill the missing regions in the photographs in order to avoid the creation of holes in the 3D surface constructed from these photographs. Note that we do not fill the holes in the 3D surface; instead, we perform the filling of holes in the source photographs that would be used to generate the 3D surface. By performing inpainting, the viewers will be provided a view of the monuments in their entirety.

The inpainting technique that we discuss here is based on the use of similar patches, i.e., exemplars to fill the missing pixels. Given a region to be inpainted, a set of exemplars is automatically searched considering a window around every pixel in this region. Here the sum of squared differences (SSD) criteria is used to determine the similarity of patches in order to search for the exemplars. The novelty of our technique lies in the use of parameters of an autoregressive (AR) model that are estimated using the non-negatively constrained least squares (NNLS) method [18]. Unlike simply copying into the target (i.e., pixels to be inpainted) values from the best-matching exemplar, as is done by many of the exemplar-based methods [21, 97, 131], we use the AR parameters in addition to the best-matching exemplar to fill the missing pixel values. For a set of candidate exemplars, the AR parameters suggest the contribution of values of neighboring pixels toward the respective center pixel of every $3 \times 3$ region in that set. A good source for filling the missing pixels (i.e., an exemplar) may not always be available in the image. In such situations, estimating a pixel value by making use of (a) knowledge of the spatial relationship of pixels and (b) information from an exemplar avoids the seam that may arise due to direct copying of pixels from the exemplar. We briefly discuss the limitations of existing approaches in Section 4.1 and

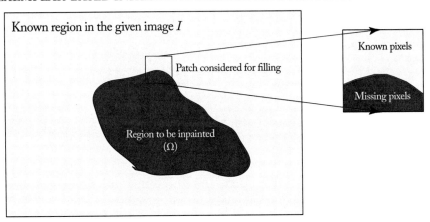

**Figure 4.1:** Setup for exemplar-based image inpainting.

describe the details of our proposed algorithm in Section 4.2. The experimental results are illustrated in Section 4.3 with the help of true images captured from the world heritage site in Hampi, Karnataka, India. Section 4.4 concludes the chapter.

## 4.1    LIMITATION OF EXISTING APPROACHES

Criminisi et al. [20, 21] proposed an inpainting technique that makes use of block replication (i.e., example patches or exemplars) to fill the missing pixels. This technique enabled the propagation of texture inside the missing regions that was earlier not possible with the methods that used partial differential equations (PDEs) [8, 80]. The exemplar-based method exploits the fact that for a small patch at the boundary of the missing region, a similar patch, i.e., an exemplar, can be found from the surrounding known region in the given image itself. As shown in Figure 4.1, the missing pixels in this patch are then filled by copying corresponding pixels from the exemplar. This method emphasizes on order of selecting the patch to be filled up, which allows the propagation of structure as well as texture. However, simply copying the pixels results in a visible seam where even the most similar exemplar is considerably dissimilar from the patch to be filled.

Another technique wherein objects/texture from either (a) the input image or (b) different images can be a source for inpainting was introduced by Pérez et al. [97]. Here the source regions provide the guidance vector field for estimating the missing pixel values. Thus, if a user supplies a region of interest to be edited and a region from another (or same) image from which information is to be transferred (guidance vector field), a seamless blending is achieved by solving for the Poisson equations. However, the guidance vector field needs to be manually selected leading to highly subjective results. Moreover, the objective function to be minimized determines the inpainted pixel values as the average of neighboring pixel values, which is not true for all images. This motivates us to use an AR model-based approach for inpainting.

## 4.2    PROPOSED APPROACH

Our work addresses the limitations mentioned in Section 4.1 of the widely used inpainting algorithms. Because of the spatial dependency of a pixel value on its neighbors, an AR model can be used to express this dependence, where a pixel value is a linear combination of values of its neighboring pixels [58]. Considering a first-order neighborhood, we make use of a set of exemplars to estimate the AR parameters. These parameters in addition to the best-matching exemplar are used as constraints for estimating the values of the missing pixels. Since the AR parameters suggest the contribution of neighbors, a non-negative least squares (NNLS) method [18, 66] is used to calculate the values of these parameters. A simple least squares (LS) method is unsuitable for this purpose as it may lead to negative values for the AR parameters, which do not correspond to the correct fractions.

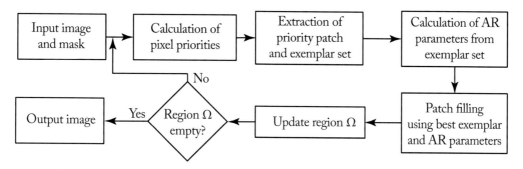

**Figure 4.2:** Proposed approach for inpainting using an AR model.

An outline of our proposed method is shown in Figure 4.2. We start with an input image $I$ and a user-defined mask that specifies the region to be inpainted or the target region $\Omega$ as shown in Figure 4.1. Let the boundary of $\Omega$ be denoted by $\delta\Omega$. Our method makes use of the approach in Criminisi et al. [21] to determine the patch priority for inpainting, wherein for every pixel $p \in I$ a confidence term $C(p)$ and data term $D(p)$ are calculated. Initially, the confidence term (initial confidence $c(p)$) is calculated as:

$$c(p) = \begin{cases} 1, & \forall p \in I - \Omega \\ 0, & \forall p \in \Omega. \end{cases} \tag{4.1}$$

Now considering a fixed size patch $\Psi_p$ around every pixel $p$, the confidence term $C(p)$ and the data term $D(p)$ are calculated as:

$$C(p) = \frac{\Sigma_{q \in \Psi_p} c(q)}{|\Psi_p|}, \quad D(p) = \frac{|\nabla I_p^{\perp} . \mathbf{n}_p|}{\alpha}, \tag{4.2}$$

where $|\Psi_p|$ is the area of the patch $\Psi_p$ of pixel $p$, $\nabla I_p^{\perp}$ is orthogonal to the gradient $\nabla I_p$ at a pixel $p$, $\mathbf{n}_p$ is unit normal to the target boundary $\delta\Omega$ and $\alpha$ is the normalization factor taken to be 255 (for grey-level images). The priority $P(p)$ associated with every pixel $p$ in $\delta\Omega$ is given by:

$$P(p) = C(p)D(p). \tag{4.3}$$

Once these priorities are calculated, the patch $\Psi_{\hat{p}}$ around the pixel $\hat{p}$ that has maximum priority is considered for filling.

Since the pixels on the boundary of the region $\Omega$ to be inpainted get more priority, the selected patch $\Psi_{\hat{p}}$ will always consist of both the known and missing pixels (as shown in Figure 4.1). We intend to make use of information from a similar patch, i.e., an exemplar to fill the missing pixels in patch $\Psi_{\hat{p}}$. An exemplar can be found by comparing the patch $\Psi_{\hat{p}}$ with a patch $\Psi_{\hat{q}}$ around every pixel $\hat{q}$ in the entire image, but as the image size increases the time taken to find an exemplar also increases. However, we observe that a patch is similar to those in its surrounding region. We therefore restrict the search for matching the patches to a large-sized window $W_{\hat{p}}$ around the patch to be filled up instead of searching the whole image. By doing so, the number of computations required to search the exemplar are considerably reduced. We measure the similarity of patches $\Psi_{\hat{p}}$ and $\Psi_{\hat{q}}$ by comparing the pixels $\hat{p}_i$ in patch $\Psi_{\hat{p}}$ that are known (i.e., $\hat{p}_i \notin \Omega$) with corresponding pixels in every patch $\Psi_{\hat{q}}$ (where $\Psi_{\hat{q}} \in W_{\hat{p}}$ and $\Psi_{\hat{q}} \cap \Omega = \phi$). The patch $\Psi_{\hat{q}}$ which gives the minimum SSD is considered as the exemplar $E_{\hat{p}}$.

Once the exemplar $E_{\hat{p}}$ is available, one may be tempted to fill the missing pixels by simply copying the corresponding pixels from the exemplar $E_{\hat{p}}$ as done in Criminisi et al. [21]. This works well if the pixel intensities in the window $W_{\hat{p}}$ do not vary much, leading to small SSD error. However, it may be noted that due to variation in illumination or contrast within the window $W_{\hat{p}}$, which is often the case for images of heritage scenes, the SSD obtained for $E_{\hat{p}}$ will be high, i.e., even the most similar patch will also be considerably different from $\Psi_{\hat{p}}$. In such cases, if pixel values are copied from the $E_{\hat{p}}$ into $\Psi_{\hat{p}}$, the modifications in $\Psi_{\hat{p}}$ do not appear to be uniform, making the seam clearly visible in the inpainted patch. In order to get better inpainting, one can think of using the pixel-neighborhood relationship while filling the missing pixels in the patch $\Psi_{\hat{p}}$.

For estimating the pixel-neighborhood relationship, we model the central pixel value of a $3 \times 3$ region to be a linear combination of the values of its first-order neighboring pixels. Let $k_t$, $k_r$, $k_b$ and $k_l$ denote the contributions of the top, right, bottom and left neighbors, respectively toward the central pixel value of a $3 \times 3$ region. We now arrange all the patches $\Psi_{\hat{q}} \in W_{\hat{p}}$ in ascending order of the SSD error and consider only the first $L$ patches to form a set $S_{\hat{p}}$ for estimating these contributions. As we know that the coefficients associated with the neighboring pixels remain constant for a small region or similar regions, for all the $3 \times 3$ regions in $S_{\hat{p}}$ we can write

$$g_p = k_t g_{q_t} + k_r g_{q_r} + k_b g_{q_b} + k_l g_{q_l}. \tag{4.4}$$

where $g_p$ is the center pixel value of every $3 \times 3$ region in $S_{\hat{p}}$ and $g_{q_t}, g_{q_r}, g_{q_b}, g_{q_l}$ are the top, right, bottom and left neighboring pixel values, respectively. The Equation (4.4) represents the equation for an AR model [38] for the pixel $g_p$. Now that the values of $g_p$ and corresponding first-order neighbors are known by considering the set $S_{\hat{p}}$, we estimate the values of $k_t$, $k_r$, $k_b$ and $k_l$ that best fit the AR model.

Using the least squares (LS) method to determine these AR parameters may result in few of them being negative. However, since the coefficients associated with neighboring pixels denote the proportion of the respective neighbor's contribution, a negative value is unacceptable. We therefore use the NNLS method [18, 66] to obtain the values of $k_t$, $k_r$, $k_b$ and $k_l$, which assures that the obtained values are non-negative. The NNLS method iteratively categorizes the constraints into *active* and *passive* sets. The constraints corresponding to a negative or zero regression coefficient are included in the active set, and the remaining constraints constitute the passive set. The solution then corresponds to the unconstrained least squares solution using only the variables corresponding to the passive set by setting the regression coefficients corresponding to the active set to zero.

The number of exemplars $L$ in the set $S_{\hat{p}}$ need to be greater than or equal to the number of AR parameters to be determined, else we are left with more numbers of unknowns to be estimated from fewer numbers of constraints. Further, as $L$ increases, we have more numbers of constraints which generalize the model leading to better estimates of AR parameters. However, if $L$ is very large, then many exemplar candidates with larger SSD values (i.e., outliers) get involved in calculating the AR parameters. As a result, the estimated values do not represent the true spatial relationship of the pixels in the patch $\Psi_{\hat{p}}$. Therefore, one has to heuristically choose the number $L$. One has to also take care that the size of patch $\Psi_{\hat{p}}$ is not very large. If a patch with a large size is considered, the size of patches in the set $S_{\hat{p}}$ (that are used to calculate the AR parameters) will also be large. Over a large region, the spatial relationship of the pixels may change and it may not be accurately represented by the AR parameters, in turn reducing the effectiveness of the algorithm. For images of heritage sites, this is of particular importance for re-creating fine artistic details over small local regions.

With the availability of the exemplar $E_{\hat{p}}$ and the pixel-neighborhood relationship in the form of AR parameters, we estimate the missing pixel in $\Psi_{\hat{p}}$ by blending the corresponding pixels from the exemplar. For this purpose we use a method derived from the work by Pérez et al. [97], which demonstrates the seamless blending of pixel values from a source region in one image into a missing region in the same or a different image, by solving for unknown pixel values using discrete Poisson equations. In their work the value of every pixel $p$ in the missing region $\Omega$ satisfies the following equation:

$$|N_p|f_p - \sum_{q \in N_p \cap \Omega} f_q = \sum_{q \in N_p \cap \delta\Omega} f_q^* + \sum_{q \in N_p} v_{pq} \tag{4.5}$$

where $f_p$ is a value of pixel $p \in \Omega$, $f_q^*$ is a value of pixel $q \in I - \Omega$, $f_q$ is a value of pixel $q \in N_p \cap \Omega$, $N_p$ is the set of neighbors of pixel $p$, $|N_p|$ is the number of pixels in the set $N_p$. Here $v_{pq}$ denotes the difference between values of a pixel $p$ and its neighboring pixel $q$ in the source region which correspond to the pixel $p$ in the missing region and its neighboring pixel

$q$, respectively. One may note that in Pérez et al. [97] both the source and missing regions are selected manually since the technique is primarily for image editing. Also, the blending is done considering all the missing pixels at once and not in a patch-based approach. In our technique, the automatically obtained exemplar corresponding to the patch selected for filling up is considered as the source region. If we re-arrange Equation (4.5) we get,

$$f_p = \frac{\sum_{q \in N_p \cap \Omega} f_q + \sum_{q \in N_p \cap \delta\Omega} f_q^* + \sum_{q \in N_p} v_{pq}}{|N_p|} \tag{4.6}$$

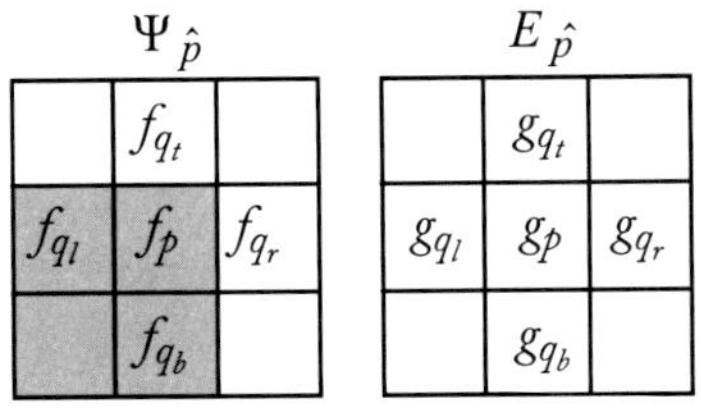

| | $\Psi_{\hat{p}}$ | | | $E_{\hat{p}}$ | |
|---|---|---|---|---|---|
| | $f_{q_t}$ | | | $g_{q_t}$ | |
| $f_{q_l}$ | $f_p$ | $f_{q_r}$ | $g_{q_l}$ | $g_p$ | $g_{q_r}$ |
| | $f_{q_b}$ | | | $g_{q_b}$ | |

**Figure 4.3:** $\Psi_{\hat{p}}$ is the patch considered for filling, $E_{\hat{p}}$ is the corresponding exemplar. The shaded region denotes the pixels in the missing region $\Omega$. $f_p$ is the value of pixel $p$ in $\Psi_{\hat{p}}$, $g_p$ is the corresponding pixel's value in the exemplar $E_{\hat{p}}$.

Considering this scenario in a patch-based approach as shown in Figure 4.3 and taking $|N_p| = 4$, the Equation (4.6) can be be written as:

$$f_p = \frac{(f_{q_b} + f_{q_l}) + (f_{q_t} + f_{q_r}) + (4g_p - (g_{q_t} + g_{q_r} + g_{q_b} + g_{q_l}))}{4} \tag{4.7}$$

where $\sum f_q = f_{q_b} + f_{q_l}$, $\sum f_q^* = f_{q_t} + f_{q_r}$ and $v_{pq} = g_p - g_q$. The above Equations (4.6) and (4.7) show that the value $f_p$ is determined by taking the average of its neighbors and by computing the average in the corresponding source patch selected manually. However, for many images, it is not true that a pixel value is an average of its neighbors. In general, pixel value can be considered as a linear combination of the first-order neighboring pixel values, and for this reason we modify the Equation (4.7) as:

$$f_p = (k_t f_{q_t} + k_r f_{q_r} + k_b f_{q_b} + k_l f_{q_l}) + g_p - (k_t g_{q_t} + k_r g_{q_r} + k_b g_{q_b} + k_l g_{q_l}). \tag{4.8}$$

Here $k_t$, $k_r$, $k_b$ and $k_l$ are the estimated AR parameters representing the contributions of each first-order neighbor as shown in Figure 4.3, while the exemplar $E_{\hat{p}}$ provides the guidance vector field. It is worth noting that when $k_t = k_r = k_b = k_l = \frac{1}{4}$, Equation (4.8) reduces to equation (4.7). The missing pixel values $f_p$ are now estimated by posing the optimization problem as follows:

$$\min \sum_{\forall f_p \in \Psi_{\hat{p}} \cap \Omega} \left\| f_p - \left( \begin{array}{c} (k_t f_{q_t} + k_r f_{q_r} + k_b f_{q_b} + k_l f_{q_l}) + \\ g_p - (k_t g_{q_t} + k_r g_{q_r} + k_b g_{q_b} + k_l g_{q_l}) \end{array} \right) \right\|^2. \tag{4.9}$$

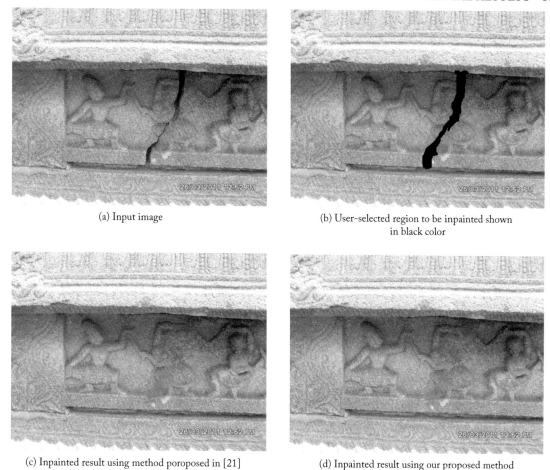

(a) Input image

(b) User-selected region to be inpainted shown
in black color

(c) Inpainted result using method poroposed in [21]

(d) Inpainted result using our proposed method

**Figure 4.4:** Result showing the inpainting of a crack in a wall carving.

Once a patch is processed, its pixels are excluded from the region to be inpainted $\Omega$. The updated missing region is then used in the next iteration and after each iteration the missing region $\Omega$ shrinks. The algorithm terminates when all the missing pixels are filled.

## 4.3    EXPERIMENTAL RESULTS

We present the results of our technique on data collected from the world heritage site, Hampi, Karnataka, India. These images were captured using a Samsung ES55 digital camera. The data consists of a number images of monuments, having both damaged and non-damaged regions. The experimental results for three such images are shown in Figures 4.4a, 4.5a and 4.6a. In all the images, fairly large cracks are visible and the aim was to restore the images as if they had no cracks at all.

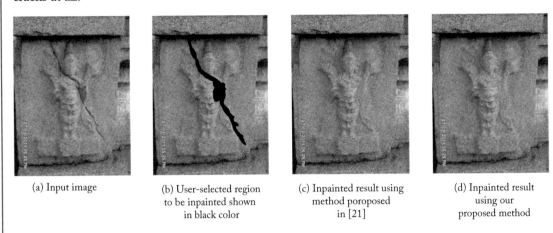

(a) Input image     (b) User-selected region to be inpainted shown in black color     (c) Inpainted result using method poroposed in [21]     (d) Inpainted result using our proposed method

**Figure 4.5:** Result showing inpainting of a long crack with varying width across a stone-work.

(a) Input image     (b) User-selected region to be inpainted shown in black color     (c) Inpainted result using method poroposed in [21]     (d) Inpainted result using our proposed method

**Figure 4.6:** Result showing inpainting to a narrow damaged portion of a statue.

We show a comparison of the results of our proposed algorithm with that of the algorithm presented in Criminisi et al. [21]. Both the algorithms are implemented in Matlab. Once an image is given as an input, the user selects the region to be inpainted, i.e., $\Omega$ that has to be filled. These regions selected by volunteers to be inpainted are shown in black color in Figures 4.4b, 4.5b and 4.6b, respectively. The results using the algorithm in Criminisi et al. [21] are shown in Figures 4.4c, 4.5c and 4.6c, respectively, and those of the proposed method are displayed in Figures 4.4d, 4.5d and 4.6d, respectively. A patch size of $9 \times 9$ was selected; the window $W_{\hat{p}}$ was chosen to be of size $37 \times 57$. In our experiment we consider the number of patches in the set $S_{\hat{p}}$ used for estimating the AR parameters to be $L = 30$.

Observe the inpainted area below the knee of the dancer in the central region in Figure 4.4c. A clear seam is visible in the inpainted area using the method proposed in Criminisi et al. [21]. On the other hand, the corresponding region shown in Figure 4.4d has been seamlessly inpainted using the proposed method. In Figure 4.5c a repetitive pattern is visible in the inpainted hand of the stone-work along with a seam on the stomach. On the contrary, in the corresponding regions shown in Figure 4.5d using the proposed method, the inpainting appears to be plausible and seamless. In the inpainted result shown in Figure 4.6, that makes use of the technique proposed in Criminisi et al. [21], a clear contrast in color can be seen inside the inpainted region. There is no such color contrast in the inpainted result of our proposed approach as seen Figure 4.6d. Here pixels inside the inpainted region exhibit an effectual blending due to which no seam is visible.

Thus, from the Figures 4.4, 4.5 and 4.6 one can notice the seam at the inpainted regions of the images in Figures 4.4c, 4.5c, 4.6c, respectively, obtained using the technique proposed in Criminisi et al. [21]. At the same time, the results obtained using our technique shown in Figures 4.4d, 4.5d and 4.6d, respectively, are seamless and plausible. Use of the proposed method to obtain seamlessly inpainted images will facilitate the generation of rich 3D models of heritage sites without the presence of holes.

## 4.4   CONCLUSION

We have presented an automatic exemplar search-based inpainting technique in this chapter. The spatial dependence of a pixel with its neighbors is used here as the cue to blend information from the exemplar into the missing pixels of the patch under consideration. Assuming the neighborhood to be of the first-order, the spatial dependence is represented using an AR model, the parameters of which are estimated from a set of candidate exemplars using the NNLS method. The proposed method avoids the direct copying of pixels from the exemplar as this results in a visible seam; instead, it uses the estimated AR parameters and the exemplar to perform a seamless blending. As seen from the experimental results, the reported results are promising. We conclude that by estimating the spatial dependence of a pixel with its neighborhood using an AR model, the damaged regions in images of monuments can be inpainted plausibly. The images of a heritage site inpainted in this manner will provide better inputs for estimating image-based 3D models.

CHAPTER 5

# Attempts to Improve Inpainting

In Chapters 2 and 3 we discussed techniques for super-resolution, while in Chapter 4, we pointed out the limitations of the existing inpainting approaches and discussed an autoregressive model-based technique to overcome the limitations. Nevertheless, the technique proposed in Chapter 4 did not deal with the problem of finding good exemplar patches for better inpainting, and instead provided a solution to improve the inpainting using the pixel-neighborhood relationship estimated from the candidate exemplars. In this chapter we discuss a few inpainting methods that we attempted in a quest to find better exemplars for improving the inpainting results. Our observations from these attempts motivated us to use a dictionary learning and compressive sensing-based approach for improving inpainting. This exercise helped us to come up with the simultaneous inpainting and super-resolution technique we will discuss later in Chapter 6.

In this chapter, we first discuss an exemplar-based multi-resolution approach in order to obtain an estimate of the pixels inside the missing region in Section 5.1. We then consider refinement, using various approaches in Sections 5.1.1–5.1.3. In Section 5.2 we also consider an inpainting approach based on curvatures in an attempt to obtain better results. We then present our observations and the conclusion we draw from these attempts in Section 5.3. One may note that the purpose of this chapter is to bridge the gap between limitations identified in Chapter 4 and our proposed solution in Chapter 6. Readers who wish to know the details of our proposed technique for simultaneous inpainting and super-resolution may directly refer to Chapter 6.

## 5.1 A MODIFIED EXEMPLAR-BASED MULTI-RESOLUTION APPROACH

An exemplar-based method like the one proposed in [21] performs inpainting in an iterative manner by processing one patch at a time. Here an exemplar is searched from outside the region to be inpainted such that it best matches the selected patch. This exemplar is then used as a source for filling the missing pixels in the selected patch. However, it may happen that there is more than one best match for the selected patch. There could also be patches that are not the best match in terms of the measure used for patch comparison, like sum of squared difference (SSD), and yet these are very similar to the patch to be filled as shown in Figure 5.1. Thus, if the first best match is considered as an exemplar as done in [21], it may not always be the best source for inpainting. This motivated us to use an additional constraint for deciding which of the candidate matches need to be considered as a source to achieve better inpainting. We then used the matching of patches in multiple resolutions as the additional constraint for exemplar selection.

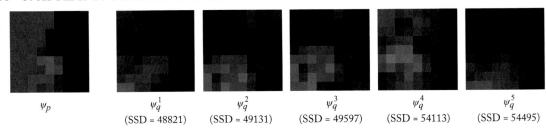

$\psi_p$  $\psi_q^1$  $\psi_q^2$  $\psi_q^3$  $\psi_q^4$  $\psi_q^5$

(SSD = 48821)  (SSD = 49131)  (SSD = 49597)  (SSD = 54113)  (SSD = 54495)

**Figure 5.1:** Matching patches using SSD. Here $\psi_p$ is the patch selected for inpainting with the missing pixels shown in red color. The patches $\psi_q^1, \ldots, \psi_q^5$ are the most similar patches to $\psi_p$ in terms of SSD. Although $\psi_q^1$ has a smaller value for SSD, we observe $\psi_q^2$ and $\psi_q^3$ to be better sources for filling the missing pixels in $\psi_p$.

In order to perform a multi-resolution matching of patches in an image $I$, consider its coarser resolution version $I_c$ obtained by blurring and downsampling by a factor of 2. If the lower resolution image is available (acquired by a real camera), it may be made use of. For the patch $\psi_p$ on the boundary of the missing region $\Omega$, we search for similar patches in the image $I$ to obtain the candidate exemplars $\psi_q^1, \ldots, \psi_q^K$. Also, we search for a similar patch $\psi_{p_c}$ in $I_c$ and consider the corresponding patch $\psi_{p_h}$ in $I$ as the finer resolution version of the patch $\psi_p$ as illustrated in Figure 5.2. Similarly, for each of these candidate exemplars we find the corresponding finer resolution versions $\Psi_{q_h}^1, \ldots, \Psi_{q_h}^K$. We then choose the exemplar $\psi_q$ using the following Equation (5.1).

$$\psi_q = \psi_q^i \quad \text{where,} \quad i = \underset{j=1 \text{ to } K}{\operatorname{argmin}} \ d(\Psi_{p_h}, \Psi_{q_h}^j), \tag{5.1}$$

and $d(\cdot, \cdot)$ is the function that calculates the patch difference for which we use the sum of squared difference. We select this patch $\psi_q$ as a source in order to obtain a better inpainted result. This method resulted in better inpainting as indicated by one of the results shown in Figure 5.3. Nevertheless, since the finer resolution patches do not represent true high-resolution versions, we consider the filled missing pixels in this way as an initial estimate and attempt to refine it using various methods discussed below.

We first considered using a larger region around the patch under consideration to improve the inpainting by copying pixels from a corresponding patch in the matched region. We then considered the relationship of a patch to be refined using the patches in its first-order neighborhood. We estimated the parameters of this relationship in the least-squares sense. We then attempted the refinement of patches by expressing their relationship with other patches in their neighborhood as a sparse representation using the compressive sensing framework. These attempted works are discussed below.

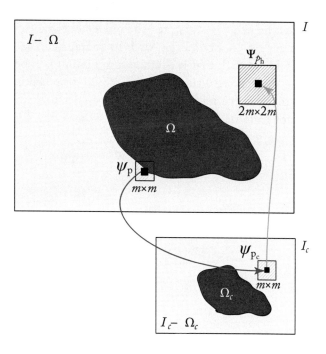

**Figure 5.2:** Estimation of the finer resolution patch $\Psi_{p_h}$ corresponding to the patch $\psi_p$ for multi-resolution patch matching.

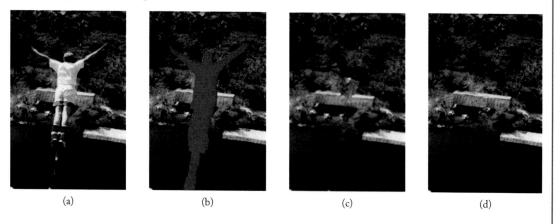

(a)                    (b)                    (c)                    (d)

**Figure 5.3:** A result demonstrating the use of exemplars obtained by matching patches at multiple resolutions on an image of a person performing bungee jump. (a) Input; (b) region to be inpainted is shown in red color; (c) inpainted result using the technique in [21]; (d) inpainted result obtained using our multi-resolution patch-matching method.

### 5.1.1  REFINEMENT BY MATCHING A LARGER REGION

Once the initial estimate is obtained, we attempted the refinement in a patch-by-patch manner by finding a match of a larger region around the patch to be refined. The larger region around the patch to be refined is selected such that it contains a minimum number of pixels that need refinement. After choosing this region, a similar region is now searched in the image containing the initially estimated pixels in the missing regions. The source for refining the patch under consideration is then selected from this region as follows.

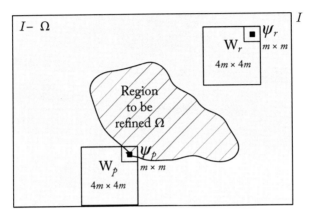

**Figure 5.4:** Selection of window $W_p$ for refinement of the patch $\psi_p$. The window $W_p$ around the patch $\psi_p$ is selected such that it has the least overlap with refinement region $\Omega$.

For a patch $\psi_p$, let $W_p$ denote the corresponding larger region. If the best match for $W_p$ is denoted by $W_r$, then the patch $\psi_r \in W_r$ as illustrated in Figure 5.4 is considered as the source for refinement. The refinement is done by replacing the inpainted pixels in $\psi_p$ with the corresponding pixels in $\psi_r$. Likewise, the other patches in the inpainted region are refined in each iteration.

The results obtained using this refinement approach are shown in Figures 5.5–5.7. From the results shown in Figures 5.5–5.7 we observe that although refined inpainted results appear better, the improvement is not significant. We therefore ventured into another method for refinement.

### 5.1.2  REFINEMENT USING THE PATCH-NEIGHBORHOOD RELATIONSHIP

In this method, the refinement of a patch was carried out by expressing it as a linear combination of its first-order neighboring patches. Since the patches to be refined contain pixel values representing the initial estimate of the missing pixels, we avoid their use in current resolution; instead, we make use of the finer resolution patches and estimate the coefficients of the linear combination in the least squares sense by assuming that the relationship is maintained across multiple resolutions. In other words, we consider the patch to be refined and its neighboring patches as low-resolution (LR) observations and estimate the coefficients of their linear combi-

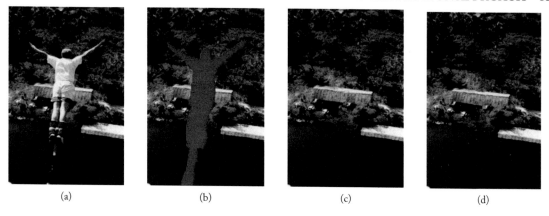

<table>
<tr><td>(a)</td><td>(b)</td><td>(c)</td><td>(d)</td></tr>
</table>

**Figure 5.5:** Result for refinement using a larger region on an image of a person performing bungee jump. (a) Input; (b) region to be inpainted is shown in red color; (c) inpainted result obtained using our multi-resolution patch-matching method; (d) refined version of (c). (Higher number of green pixels are seen on the roof of the house after refinement.)

<table>
<tr><td>(a)</td><td>(b)</td><td>(c)</td><td>(d)</td></tr>
</table>

**Figure 5.6:** Result for refinement using a larger region on an image of toy1. (a) Input; (b) region to be inpainted is shown in red color; (c) inpainted result obtained using our multi-resolution patch-matching method; (d) refined version of (c). (Inconsistent patches can be observed after refinement.)

nation by using the corresponding finer resolution patches obtained from our multi-resolution patch-matching method. The estimated coefficients are then used along with the neighborhood patches to refine the patch under consideration.

Here for a patch $\psi_p$ to be refined, we obtain the corresponding finer resolution patch $\Psi_{p_h}$ using the method illustrated by Figure 5.2. Similarly, for every first-order neighboring patch of $\psi_p$, we can obtain the corresponding HR patches. Let the first-order neighboring patches be denoted by $\psi_{p^1}, \ldots, \psi_{p^4}$ and their respective finer resolution patches be $\Psi_{p_h^1}, \ldots, \Psi_{p_h^4}$. We now express each pixel in the patch $\Psi_{p_h}$ as the linear combination of the corresponding pixels in the patches $\Psi_{p_h^1}, \ldots, \Psi_{p_h^4}$. The coefficients of this linear combination are obtained in the least-square sense. Using these coefficients and the patches $\psi_{p^1}, \ldots, \psi_{p^4}$, we refine the pixels in the patch $\psi_p$.

(a)          (b)          (c)          (d)

**Figure 5.7:** Result for refinement using a larger region on an image of toy2. (a) Input; (b) region to be inpainted is shown in red color; (c) inpainted result obtained using our multi-resolution patch-matching method; (d) refined version of (c). (Seam is visible after in the refined region.)

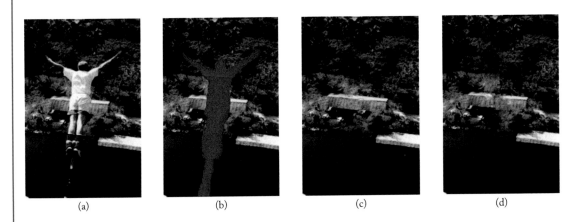

(a)          (b)          (c)          (d)

**Figure 5.8:** Result for refinement using the patch-neighborhood relationship on an image of a person performing bungee jump. (a) Input; (b) region to be inpainted is shown in red color; (c) inpainted result obtained using our multi-resolution patch-matching method; (d) refined version of (c). (Seam is visible on the boundary of the refined patches.)

The results for this refinement method are shown in Figures 5.8–5.10. From these results, we observe no improvement after refining the region. In fact, the results are visually not plausible as the seam is visible on the boundary of refined patches. This left us to aim at yet another method for refinement based on the compressive sensing framework as discussed below.

### 5.1.3   REFINEMENT USING COMPRESSIVE SENSING FRAMEWORK

In this method, the refinement of the region in a patch-by-patch manner is done using the compressive sensing (CS) framework [13]. Here we consider a window, as illustrated in Figure 5.4, around the patch to be refined and express this patch as a sparse combination of the other patches in the window.

(a)        (b)        (c)        (d)

**Figure 5.9:** Result for refinement using the patch-neighborhood relationship on an image of toy1. (a) Input; (b) region to be inpainted is shown in red color; (c) inpainted result obtained using our multi-resolution patch-matching method; (d) refined version of (c). (Seam is visible on the boundary of the refined patches.)

(a)        (b)        (c)        (d)

**Figure 5.10:** Result for refinement using the patch-neighborhood relationship on an image of toy2. (a) Input; (b) region to be inpainted is shown in red color; (c) inpainted result obtained using our multi-resolution patch-matching method; (d) refined version of (c). (Seam is visible on the boundary of the refined patches.)

Let $W_p$ denote the window around the patch $\psi_p$ to be refined and let $\psi_p^1, \ldots, \psi_p^L$ denote the patches except $\psi_p$ in this window. Consider the exemplar $\psi_q$ obtained using the multi-resolution patch-matching method that is similar to the patch $\psi_p$ and a window $W_q$ around it consisting of patches $\psi_q^1, \ldots, \psi_q^L$. We now perform lexicographical ordering of the patches $\psi_q^1, \ldots, \psi_q^L$ and construct a dictionary $D_q$ using these as its columns. The lexicographically ordered version $y_q$ of the exemplar $\psi_q$ is then expressed as a spare combination of the columns in the dictionary $D_q$. The coefficients $\alpha$ of this sparse combination are obtained, posing the problem in a CS framework as,

$$\min \|\alpha\|_{l_1} \quad \text{such that} \quad y_q = D_q * \alpha, \tag{5.2}$$

where $\|\alpha\|_{l_1}$ corresponds to $\sum_{j=1}^{L} |\alpha_j|^1$. Minimization is carried out using standard optimization tools [12]. Once the coefficients are estimated, the dictionary $D_p$ constructed by using $\psi_p^1, \ldots, \psi_p^L$ is used to reconstruct $\psi_p$, giving us the refined patch. The lexicographically ordered version $y_p$ of $\psi_p$ is obtained by using $y_p = D_p * \alpha$.

In order to illustrate this approach, we conducted a few experiments where the window size was set to four times the size of the patches. The result obtained are shown below in Figures 5.11–

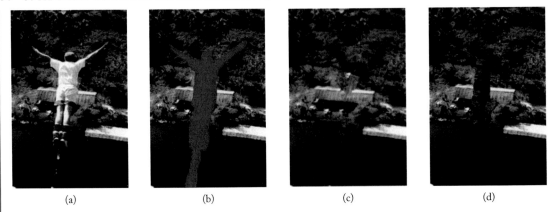

(a)                    (b)                    (c)                    (d)

**Figure 5.11:** Result for refinement using the compressive sensing framework on an image of a person performing bungee jump. (a) Input; (b) region to be inpainted is shown in red color; (c) inpainted result obtained using our multi-resolution patch-matching method; (d) refined version of (c). (The result appears implausible after refinement.)

(a)                    (b)                    (c)                    (d)

**Figure 5.12:** Result for refinement using the compressive sensing framework on an image of toy1. (a) Input; (b) region to be inpainted is shown in red color; (c) inpainted result obtained using our multi-resolution patch-matching method; (d) refined version of (c). (Refined region appears unrealistic.)

5.13. From the results, we observe that this refinement method does not help in improving the inpainting. On the contrary, it worsens the already inpainted region that we use as our initial estimate. We therefore attempted an altogether different approach that is based on curvatures. This approach is discussed in the following Section 5.2.

## 5.2   CURVATURE-BASED APPROACH FOR INPAINTING

The refinement methods for the inpainting approach discussed in the previous Section 5.1 were unable to generate seamlessly inpainted results. We therefore explored a different approach for inpainting based on curvatures. Note that this an independent approach for inpainting and not a method for refinement. In order to have a plausibly inpainted image, the structures arriving at

   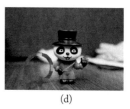

| (a) | (b) | (c) | (d) |

**Figure 5.13:** Result for refinement using the compressive sensing framework on an image of toy2. (a) Input; (b) region to be inpainted is shown in red color; (c) inpainted result obtained using our multi-resolution patch-matching method; (d) refined version of (c). (The refined result appears poor in comparison to the initial estimate.)

the boundary of the regions to be inpainted need to smoothly propagate inside these regions. The curvature of the contours of these structures could be constrained to achieve this smooth propagation that helps in seamless inpainting. With this motivation, in this approach we formulate a cost function such that the difference between the curvatures of the second-order neighboring pixels is minimized. For an image $I$, the curvature $\kappa$ evaluated at pixels with coordinates $(x, y)$ is given by Sapiro [107],

$$\kappa = \frac{I_x I_{yy} - I_y I_{xx}}{(I_x^2 + I_y^2)^{\frac{3}{2}}}. \tag{5.3}$$

Let $\Omega$ denote the region to be inpainted and a pixel $p \in \Omega$ has the coordinates $(x, y)$. Also, let $N_p$ be the neighborhood of $p$. In order to estimate the pixel intensity $I(p)$, we formulate the cost function as,

$$\min \sum_{p \in \Omega} \left[ \sum_{q \in N_p} (\kappa(p) - \kappa(q))^2 + \lambda \sum_{q \in N_p} (I(p) - I(q))^2 \right], \tag{5.4}$$

where the first term is the data term, the second term is for maintaining the smoothness and $\lambda$ is a constant.

Solving the above equation simultaneously for all the missing pixels $p \in \Omega$ is difficult as this involves non-linear terms. We therefore estimate the intensity of the missing pixels at the boundary of $\Omega$ first and then move inward to the complete region $\Omega$ in subsequent iterations.

If the boundary is denoted by $\delta\Omega$ then the cost function for estimating value of a pixel $p \in \delta\Omega$ is given as follows,

$$\min \sum_{q \in N_p \cap (I - \Omega)} (\kappa(q) - \kappa(q_n))^2 + \lambda \sum_{q \in N_p \cap (I - \Omega)} (I(p) - I(q))^2, \tag{5.5}$$

where $q_n$ is the neighbor of $q$ such that $p - q = q - q_n$. Thus the pixel $q_n$ is farther away from pixel $p$ in the same direction as that of the pixel $q$ from pixel $p$. Here both $q$ and its neighbor

$q_n$ are outside the region to be inpainted so that the curvature at these pixels can be calculated. We make use of curvature values at these pixels for estimating the value of the missing pixel $p$ in Equation (5.5).

Now the missing pixel $p$ can be in the second-order neighborhood of more than one pixel $q$ which is outside the region to be inpainted. Considering the combined effect in all the eight scenarios for which the missing pixel $p$ is in the second-order neighborhood of a known pixel $q$, the missing pixel value $I(p)$ is estimated by differentiating the Equation (5.5) with respect to the unknown $I(p)$ and equating it to zero. Once this is done $I(p)$ can be obtained as,

$$I(x, y) = I(p) = \frac{\left[ \sum_{i=3,4,7,8} b_i D_i^2 (c_i - d_i \kappa_i) - \sum_{i=1,2,4,5} b_i D_i^2 (c_i - d_i \kappa_i) + A\lambda D^2 \right]}{\sum_{i=1}^{8} (b_i D_i)^2 + 8\lambda D^2}, \quad (5.6)$$

where, $i = 1, \ldots, 8$ indicate the second-order neighbors such that,

$$A = \sum_{i=1}^{8} a_i, \quad D = \prod_{i=1}^{8} d_i, \quad D_i = \prod_{j=1, j \neq i}^{8} d_j,$$

$$a_1 = I(x, y - 1), \quad b_1 = (a_1 - I(x - 1, y - 1)), \quad \kappa_1 = \kappa(x, y - 2),$$
$$c_1 = b_1(I(x, y - 2) - 2a_1) - (a_1 - I(x, y - 2))(I(x - 1, y - 1) + I(x + 1, y - 1) - 2a_1),$$
$$d_1 = (b_1^2 + (a_1 - I(x, y - 2))^2)^{\frac{3}{2}},$$

$$a_2 = I(x - 1, y - 1), \quad b_2 = (a_2 - I(x - 2, y)), \quad \kappa_2 = \kappa(x - 2, y - 2),$$
$$c_2 = b_2(I(x - 2, y - 2) - 2a_2) - (a_2 - I(x - 2, y - 2))(I(x, y - 2) + I(x - 2, y) - 2a_2),$$
$$d_2 = (b_2^2 + (a_2 - I(x - 2, y - 2))^2)^{\frac{3}{2}},$$

$$a_3 = I(x - 1, y), \quad b_3 = (a_3 - I(x - 1, y - 1)), \quad \kappa_3 = \kappa(x - 2, y),$$
$$c_3 = (a_3 - I(x - 2, y))(I(x - 1, y - 1) + I(x - 1, y + 1) - 2a_3) - b_3(I(x - 2, y) - 2a_3),$$
$$d_3 = (b_3^2 + (a_3 - I(x - 2, y))^2)^{\frac{3}{2}},$$

$$a_4 = I(x - 1, y + 1), \quad b_4 = (a_4 - I(x - 2, y)), \quad \kappa_4 = \kappa(x - 2, y + 2),$$
$$c_4 = (a_4 - I(x - 2, y + 2))(I(x, y + 2) + I(x - 2, y) - 2a_4) - b_4(I(x - 2, y + 2) - 2a_4),$$
$$d_4 = (b_4^2 + (a_4 - I(x - 2, y + 2))^2)^{\frac{3}{2}},$$

$$a_5 = I(x, y + 1), \quad b_5 = (I(x + 1, y + 1) - a_5), \quad \kappa_5 = \kappa(x, y + 2),$$
$$c_5 = b_5(I(x, y + 2) - 2a_5) - (I(x, y + 2) - a_5)(I(x + 1, y + 1) + I(x - 1, y + 1) - 2a_5),$$
$$d_5 = (b_5^2 + (I(x, y + 2) - a_5)^2)^{\frac{3}{2}},$$

$$a_6 = I(x+1, y+1), \quad b_6 = (I(x+2, y) - a_6), \quad \kappa_6 = \kappa(x+2, y+2),$$
$$c_6 = b_b(I(x+2, y+2) - 2a_6) - (I(x+2, y+2) - a_6)(I(x+2, y) + I(x, y+2) - 2a_6),$$
$$d_6 = (b_6^2 + (I(x+2, y+2) - a_6)^2)^{\frac{3}{2}},$$

$$a_7 = I(x+1, y), \quad b_7 = (I(x+1, y+1) - a_7), \quad \kappa_7 = \kappa(x+2, y),$$
$$c_7 = (I(x+2, y) - a_7)(I(x+1, y-1) + I(x+1, y+1) - 2a_7) - b_7(I(x+2, y) - 2a_7),$$
$$d_7 = (b_7^2 + (I(x+2, y) - a_7)^2)^{\frac{3}{2}},$$

$$a_8 = I(x+1, y-1), \quad b_8 = (I(x+2, y) - a_8), \quad \kappa_8 = \kappa(x+2, y-2),$$
$$c_8 = (I(x+2, y-2) - a_8)(I(x+2, y) + I(x, y-2) - 2a_8) - b_8(I(x+2, y-2) - 2a_8),$$
$$d_8 = (b_8^2 + (I(x+2, y-2) - a_8)^2)^{\frac{3}{2}}.$$

Results obtained using this method are shown in Figures 5.15–5.16. These results indicate that the method performs visually better for images with a blurred object or background as shown in Figures 5.15c–5.16c. Yet they exhibit a seam near the central part of the inpainted regions. Moreover, this technique fails to synthesize texture in large missing regions as seen in Figure 5.14c, which is also a limitation of inpainting approaches based on level lines [8, 80].

## 5.3   OBSERVATIONS AND CONCLUSION

The search for better source patches for improving the inpainting using our multi-resolution-based matching of patches provided encouraging results as illustrated in Section 5.1. This indicates that the idea of searching for better exemplars indeed has the potential to yield superior inpainted results. In this method, the finer resolution patches that we obtained did not represent the true high-resolution (HR) version of the low-resolution (LR) patches. We therefore considered the obtained inpainted results as the initial estimates and applied refinement, which however did not improve the inpainting.

In our first method for refinement we considered matching for a larger window surrounding the patch to be refined. Here we observed that good matches for the larger windows were often not available, due to which the method could not provide improved inpainted results after the refinement operation. Our second method considered the refinement of a patch by using the relationship estimated between the exemplar and its neighboring patches. We realized that since the exemplar selection was not performed by comparing true HR patches, the selected exemplars were in general not the best sources and do not provide the patch-neighborhood relationship required for refinement. Our method, based on the CS framework, considered performing the refinement of a patch by constructing dictionaries using patches in a window around it. We observed that constructed dictionaries usually did not contain patches similar to the one being refined, due to which good sparse representation was not obtained. As a result, the refinement step worsened the results.

(a)                    (b)                    (c)

Figure 5.14: Result using curvature-based approach for inpainting on an image of a person performing bungee jump. (a) Input image; (b) region to be inpainted is shown in red color; (c) inpainted result. (Inpainted result is blurred and looks implausible.)

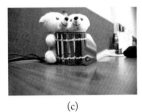

(a)                    (b)                    (c)

Figure 5.15: Result using curvature-based approach for inpainting on an image of toy1. (a) Input image; (b) region to be inpainted is shown in red color; (c) inpainted result. (Inpainted result is blurred.)

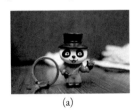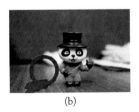

(a)                    (b)                    (c)

Figure 5.16: Result using curvature-based approach for inpainting on an image of toy2. (a) Input image; (b) region to be inpainted is shown in red color; (c) inpainted result. (Inpainted result is blurred and seam is visible in the central area of the inpainted region.)

From these experiments we realized that, in order to strongly impose the constraint of multi-resolution matching of patches, the estimation of true HR patches is required in place of the refinement methods. The method used for estimating the finer resolution patches did not address the issue of unavailability of a good matching patch in the coarser resolution image. Hence we must address this issue in order to obtain true HR patches for comparison. We also realized that the use of information from neighboring patches is not sufficient for refinement. Moreover, for obtaining good sparse representations, the dictionaries need to be constructed using a large number of patches similar to the one being refined. We also learned that the refinement step could be avoided altogether and the dictionary-based approach for sparse representation could be used to estimate a patch whenever a good match was not available in the coarser resolution image for multi-resolution patch matching. This is especially helpful for images of heritage sites since they contain artistic works that usually have self-similar patches across varying spatial resolutions. By considering image representative patches, the dictionaries for both LR and HR patches can be constructed to obtain the HR of any patch in the given image.

We conclude that in the process of searching for a better exemplar, we can thus obtain true HR patches for all the LR patches in the given image. This not only enables better inpainting at the given spatial resolution but also at a higher resolution simultaneously. The observations and conclusions drawn from this chapter enabled us to come up with the simultaneous inpainting and super-resolution technique that we discuss next in Chapter 6.

The curvature-based inpainting approach discussed in Section 5.2 introduces blur and can be used for inpainting regions having blurred objects/background. However, we have seen that the exemplar-based approaches better synthesize the texture inside large missing regions, whereas the curvature-based approach led to poor inpainting results. We therefore do not proceed further with this approach toward the development of a better inpainting technique.

# Simultaneous Inpainting and Super-resolution

In Chapter 4 we discussed an image inpainting technique that fills the missing pixels by using the pixel-neighborhood relationship in exemplars, while in Chapter 5 we discussed various attempts to improve inpainting. Based on our observations from Chapter 5, in this chapter we discuss a method that not only inpaints the given missing region but also performs super-resolution (SR) simultaneously. The past two decades have seen significant advancement in the techniques for inpainting and super-resolution. Although both problems involve the searching and processing of similar patches for estimating the unknown pixel values, the two problems have been addressed independently. As already explained in Chapters 2–4, both the inpainting and super-resolution can be used as preliminary steps for creating 3D models in applications like immersive walkthrough systems. However, the usual practice is to solve these two problems independently in a pipelined manner, i.e., first inpaint and then super-resolve. This chapter provides a unified framework to perform simultaneous inpainting and super-resolution.

In this approach, we construct dictionaries of image-representative low- and high-resolution (LR-HR) patch pairs from the known regions in the test image and its coarser resolution. The inpainting of missing pixels is then performed using exemplars that are found by comparing patch details at a finer resolution. These patches represent the higher resolution patches in the missing regions and we obtain them from the constructed dictionaries by using the self-learning approach as discussed in Chapter 2. Here the advantages when compared to other exemplar-based inpainting techniques are (a) the constraint in the form of finer resolution patch-matching results in good exemplars and better inpainting, and (b) inpainting is obtained not only in the given spatial resolution but also at higher resolution, leading to super-resolution inpainting. In other words, we obtain SR as a consequence of inpainting, thus reducing the number of computations when compared to performing these operations independently.

Note that our approach performs super-resolution without introducing blur or artefacts indicating better inpainting at the given resolution. Also, our method does not use any kind of regularization as used by most of the super-resolution approaches [101, 137]. We once again emphasize that the primary goal here is to obtain better inpainting. The super-resolution is obtained as a by-product since we use a constraint that helps in finding a better source for inpainting. One of the results of this method for a natural image is shown in Figure 6.1.

**Figure 6.1:** Simultaneous inpainting and super-resolution: (a) input; (b) region to be inpainted marked in red color; (c) inpainting using planar structure guidance [51]; (d) inpainting using proposed method indicated by a region inside the rectangular area with a yellow boundary; (e) simultaneously inpainted and super-resolved image (by a factor of 2) using the proposed method with known regions upsampled using bicubic interpolation; (f)–(h) expanded versions after upsampling (inside region marked by the rectangular area with a yellow boundary in (d)) using various approaches viz. (f) bicubic interpolation, (g) Glasner et al.'s method [42] and (h) proposed method for super-resolution.

The contents of this chapter are organized as follows. In Section 6.1 we discuss the need for comparing patches at finer resolution. The proposed approach is discussed in Section 6.2. The efficacy of this method in comparison to state-of-the-art methods is illustrated by showing results on natural images in Section 6.3 where we also present the results for images captured at heritage sites. The chapter ends with the conclusion in Section 6.4.

## 6.1   NEED FOR PATCH COMPARISON AT FINER RESOLUTION

Before we enter into the discussion of our approach for simultaneous inpainting and super-resolution, we would like to point out the need for comparing patches at a finer resolution. Natural images including those of heritage scenes usually contain many self-similar patches. This cue has been used effectively by exemplar-based inpainting methods [21, 131], where search is done for the region to be filled. However, when similar patches are unavailable, the inpainting may not be seamless, resulting in graphical garbage. Even when similar patches are available, the best match may not always be a good source for inpainting. The reason is that the patch to be filled has too few known pixels to obtain a reliable match. One may increase the patch size to have more number of known pixels. However, we may not find good matches for larger patches, and therefore the inpainted regions look implausible.

In exemplar-based inpainting approaches, patch matching is done by discarding the missing pixels. Due to this, it may happen that a better source patch for inpainting could be found among

**Table 6.1:** Description of the symbols used in this chapter

| Symbols | Meaning |
| --- | --- |
| $I_0, I_{-1}$ | Input image and its coarser resolution. |
| $\Omega_0, \Omega_{-1}$ | Region to be inpainted in the input image $I_0$ and corresponding region in $I_{-1}$. |
| $y_p$ | Patch of size $m \times m$ around a pixel $p \in I_0$. |
| $y_p^u$ | Unknown missing pixels in the patch $y_p$ that are to be inpainted, i.e., $y_p^u \in \Omega_0$. |
| $y_p^k$ | Known pixels in the patch $y_p$, i.e., $y_p^k \in I_0 - \Omega_0$. |
| $K$ | Number of candidate exemplars. |
| $y_{q_1}, \ldots, y_{q_K}$ | Candidate exemplars corresponding to the patch $y_p$. |
| $N$ | Number of patch pairs used for constructing the LR-HR dictionaries. |
| $D_{LR}$ | Dictionary of low-resolution patches. Dimension: $m^2 \times N$. |
| $D_{HR}$ | Dictionary of high-resolution patches. Dimension: $4\,m^2 \times N$. |
| $D_{LR_p}^k$ | Dictionary of low-resolution patches containing only those rows that correspond to the known pixels $y_p^k$. Dimension: $\lvert y_p^k \rvert \times N$. |
| $\alpha$ | Sparse vector of size $N \times 1$. |
| $Y_p$ | HR patch of size $2m \times 2m$ corresponding to LR patch $y_p$. |
| $Y_p^u, Y_p^k$ | HR pixels in patch $Y_p$ that correspond to the pixels $y_p^u$ and $y_p^k$, respectively, in the LR patch $y_p$. |
| $Y_{q_1}, \ldots, Y_{q_K}$ | HR patches corresponding to the candidate exemplars $y_{q_1}, \ldots, y_{q_K}$. |
| $Y_q$ | Best match for $Y_p$ among $Y_{q_1}, \ldots, Y_{q_K}$. |
| $H_p$ | Final inpainted HR patch corresponding to the LR patch $y_p$. |
| $L_p$ | Inpainted version of the LR patch $y_p$. |

the patches other than the best match. It is desirable to consider these patches as candidate sources for inpainting without discarding them. Intuitively, by performing a detailed assessment of the patches to be filled, one can confidently determine which among the candidates is a better source for inpainting. In other words, if the high-resolution (HR), i.e., finer resolution of the patches are made available, they can be used to find a reliable match, which is a better exemplar for inpainting. Considering this intuition as a cue we now discuss our proposed approach.

## 6.2    PROPOSED APPROACH

The symbols used throughout this chapter are briefly described in Table 6.1 and our proposed approach is summarized in Table 6.2. The details of our approach are as follows.

**Table 6.2:** Summary of our approach for simultaneous inpainting and super-resolution

| | |
|---|---|
| 1 | Construct LR-HR pair dictionaries using the known regions in $I_0$ and $I_{-1}$. |
| 2 | Select highest priority patch $y_p = y_p^k \cup y_p^u$ for inpainting using method in [21]. Here $y_p^k \in I_0$ and $y_p^u \in \Omega_0$. |
| 3 | Search for candidate sources (exemplars) $y_{q_1}, \dots, y_{q_K}$ in $I_0$. |
| 4 | Self-learn HR patches $Y_{q_1}, \dots, Y_{q_K}$ and $Y_p$ using the constructed dictionaries:<br>(a) Obtain $Y_{q_1}, \dots, Y_{q_K}$ corresponding to $y_{q_1}, \dots, y_{q_K}$.<br>(b) Estimate $Y_p$ corresponding to $y_p$. |
| 5 | Find best exemplar $Y_q$ in HR by comparing $Y_p$ with $Y_{q_1}, \dots, Y_{q_K}$. |
| 6 | Obtain final inpainted HR patch $H_p$ using $Y_p$ and $Y_q$. |
| 7 | Obtain inpainted LR patch $L_p$ from $H_p$ using transformation estimated from the constructed dictionaries and update $\Omega_0$. |
| 8 | Repeat steps 2–7 till all patches in $\Omega_0$ are inpainted. |

Given an image $I_0$ that has a region $\Omega_0$ to be inpainted, we obtain the coarser resolution image $I_{-1}$ by blurring and downsampling $I_0$ as done in Glasner et al. [42]. Let $\Omega_{-1}$ denote the missing region in $I_{-1}$, which corresponds to $\Omega_0$ as shown in Figure 6.2.

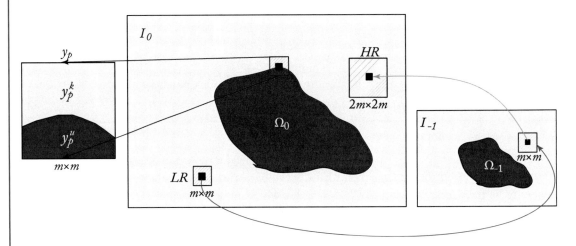

**Figure 6.2:** Finding LR-HR patch pairs using given image $I_0$ and its coarser resolution $I_{-1}$.

We now select a patch $y_p$ of size $m \times m$ around a pixel $p$ on the boundary of $\Omega_0$ for filling, based on a priority order that depends on the presence of structure and proportion of known pixels in the patch $y_p$. For calculating the priority we use the method proposed in Criminisi et al. [21]

that was explained earlier in Section 4.2. Let $y_p^k$ and $y_p^u$ denote the known and the unknown pixels in $y_p$. The patch $y_p$ is then compared with every $m \times m$ sized patch in the known region $I_0 - \Omega_0$, using sum of squared difference (SSD) by considering only the pixels corresponding to $y_p^k$. We then obtain $K$ best matches denoted as $y_{q_1}, \ldots, y_{q_K}$, representing the candidate exemplars. The exemplar-based methods [21, 131] use $K = 1$ to obtain the best match, whereas our method considers more candidate matches by setting $K > 1$ in order to find a better exemplar. These patches are then used in obtaining HR patches.

Khatri and Joshi [62] have shown that HR details can be self-learned from the given image and its single coarser resolution. Drawing inspiration from Khatri and Joshi [62], the proposed method estimates the HR details even for patches with missing pixels. To do this we first find LR-HR matches for known regions over the entire image. Consider an LR patch of size $m \times m$ in the known region $I_0 - \Omega_0$. We can obtain the corresponding $2m \times 2m$ sized HR patch in the same resolution by considering the coarser resolution $I_{-1}$ as illustrated in Figure 6.2. Although not all LR patches can find a good match in the coarser resolution, we use this methodology to create dictionaries of image-representative LR-HR patch pairs, with the help of which a good match can be estimated for any LR patch in the known region. We also learn the HR of an LR patch $y_p$ with missing pixels (i.e., $y_p^u \in \Omega_0$) by making use of these LR-HR patch pairs. Simultaneous inpainting and SR of the missing pixels is then performed by refining the estimated HR of $y_p$, using HR of the best candidate among $y_{q_1}, \ldots, y_{q_K}$ and an LR-HR relationship learned from the known region. Thus, we make use of self-learning while obtaining the HR patches of inpainting region, which are then used to obtain the corresponding inpainted LR patches. In what follows we provide the details of (a) constructing image-representative LR-HR dictionaries, (b) estimation of HR patches and (c) simultaneous inpainting and SR of missing pixels.

## 6.2.1    CONSTRUCTING IMAGE-REPRESENTATIVE LR-HR DICTIONARIES

To obtain the image-representative LR-HR patch pairs, we consider every $m \times m$ sized patch in the known region $I_0 - \Omega_0$. For each of these patches we find the best match by searching for similar patches in $I_{-1} - \Omega_{-1}$. We then get the corresponding HR in $I_0 - \Omega_0$. Here every LR patch will be mapped to exactly one HR patch. However, an HR patch may be mapped by many LR patches when the LR patches are similar, as seen in Figure 6.3 where $LR_1$–$LR_3$ patches map to a single HR patch in $I_0$. We then create a plot of HR patches vs. the number of LR patches that each HR patch is mapped to. The HR patches that are highly mapped indicate repetitiveness of the LR patches and are therefore appropriate for representing the image patches. On the other hand, the HR patches having fewer LR mappings are less likely to represent the patches inside the region to be filled. Such patches are therefore discarded.

The highly mapped HR patches form the HR dictionary $D_{HR}$ of size $4m^2 \times N$ and the corresponding $m \times m$ sized patches in $I_{-1} - \Omega_{-1}$ form the LR dictionary $D_{LR}$ of size $m^2 \times N$.

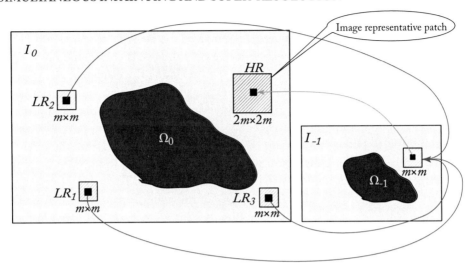

**Figure 6.3:** Example of an image representative patch.

Here $N$ is the number of highly mapped patches such that $N >> 4m^2$. Note that the dictionaries constructed in this way do not have LR-HR pairs for every patch in the known region of $I_0$.

## 6.2.2  ESTIMATION OF HR PATCHES

For an LR patch whose match is directly available in the LR dictionary, the corresponding patch in the HR dictionary is the required HR patch. For other LR patches we estimate a good match using a linear combination of few patches in the LR dictionary. When a signal is known to be sparse, the compressive sensing (CS) theory [13] provides a method to obtain the sparse representation. In our case, an LR patch $y$ whose HR version needs to be estimated, can be sparsely represented using the LR dictionary $D_{LR}$ such that:

$$y = D_{LR} * \alpha, \tag{6.1}$$

where $\alpha$ is a sparse vector of size $N \times 1$ and $y$ represents the lexicographically ordered LR patch of size $m^2 \times 1$. The sparse vector $\alpha$ is obtained by posing the problem as:

$$\min \|\alpha\|_{l_1}, \quad \text{subject to} \quad y = D_{LR} * \alpha, \tag{6.2}$$

where $\|\alpha\|_{l_1}$ corresponds to $\sum_{j=1}^{N} |\alpha_j|^1$ which is minimized using standard optimization tools [12]. In this way, we obtain good matches from the already available LR dictionary itself. Assuming the LR-HR patch pairs to have the same sparseness and using the estimated sparse coefficients ($\alpha$), the corresponding HR patch $Y$ of size $4m^2 \times 1$ is obtained as follows:

$$Y = D_{HR} * \alpha, \tag{6.3}$$

where $D_{HR}$ denotes the HR dictionary. The pixels in $Y$ are rearranged to get a patch of size $2m \times 2m$ by reversing the operation that was used to obtain the lexicographical ordering. This procedure is used to obtain the HR patches $Y_{q_1}, \ldots, Y_{q_K}$ corresponding to the $K$ candidate source patches $y_{q_1}, \ldots, y_{q_K}$ by replacing $y = y_{q_i}$, $\alpha = \alpha_{q_i}$ and $Y = Y_{q_i}$ for $i = 1, \ldots, K$ in the above equations.

The patch $y_p$ that needs to be inpainted has missing pixels $y_p^u$. Therefore, one cannot directly obtain the corresponding HR patch. However, the known pixels $y_p^k$ can be represented using a reduced LR dictionary $D_{LR_p}^k$ which consists of only those rows in $D_{LR}$ that correspond to the pixels $y_p^k$ depending on which of the pixels in $y_p$ are missing. Here $D_{LR_p}^k$ is of size $|y_p^k| \times N$ where $|y_p^k|$ denotes the number of known pixels in $y_p$. We now obtain $Y_p$ corresponding to $y_p$ by replacing $y = y_p^k$, $\alpha = \alpha_p$, $D_{LR} = D_{LR_p}^k$ and $Y = Y_p$ in Equations (6.1)–(6.3). Note that, in order to obtain $Y_p$ we use the complete HR dictionary $D_{HR}$ of size $4m^2 \times N$ and hence $Y_p$ has the size of $2m \times 2m$, i.e., it has no missing pixels. Since $Y_p$ is obtained by considering only the known pixels $y_p^k \in y_p$ and the corresponding dictionary $D_{LR}^k$, the pixels $Y_p^k$ that correspond to $y_p^k$ represent true HR pixels. Likewise, the HR pixels $Y_p^u$ that correspond to $y_p^u$ provide a better approximation to the missing HR pixels due to the use of many similar and representative patches.

## 6.2.3   SIMULTANEOUS INPAINTING AND SR OF MISSING PIXELS

The final HR patch selection for missing regions is done using $Y_p$ and $Y_{q_1}, \ldots, Y_{q_K}$ as follows. We compare each of the HR patches $Y_{q_1}, \ldots, Y_{q_K}$ with $Y_p$ and choose the one that has minimum SSD as $Y_q$. As the pixels in $Y_p^u$ represent an approximate but not true HR version of the missing pixels we replace them with those in $Y_q$ in which all pixels represent true HR. The resulting patch $H_p$ is a final HR patch which is then used to obtain the LR patch $L_p$, representing the inpainted version of the patch $y_p$.

In order to obtain $L_p$ from $H_p$ we need the HR to LR transformation. In our case, blurring and downsampling is used to obtain coarser resolution $I_{-1}$ from $I_0$ as done in Glasner et al. [42]. Hence the same operation is used to obtain $L_p$ from $H_p$. However, if the point spread function (PSF) of the camera is available, one can use it and perform downsampling to obtain the coarser resolution patches. Alternatively, if one uses $I_{-1}$ that is captured using a camera, then the HR to LR transformation can be estimated from the available dictionaries having true LR-HR patch pairs to get $L_p$ from $H_p$. Once the LR-HR dictionary pair is available we can model each LR pixel $lr_i$ as a linear combination of four HR pixels $hr_i^{00}, hr_i^{01}, hr_i^{10}$, and $hr_i^{11}$ as follows:

$$lr_i = [hr_i^{00} \ hr_i^{01} \ hr_i^{10} \ hr_i^{11}][a_{00} \ a_{01} \ a_{10} \ a_{11}]^T, \tag{6.4}$$

where $a_{00}, a_{01}, a_{10}$ and $a_{11}$ are the coefficients of the linear combination. Using the pixels in the LR-HR pair dictionaries in Equation (6.4) these coefficients can be estimated in the least-squares sense. We can then obtain $L_p$ from $H_p$ by making use of the estimated coefficients.

We now have both LR and corresponding HR patches which are inpainted. The patch $H_p$ is now placed appropriately in the upsampled image to obtain SR of the inpainted region. This process is repeated to inpaint the entire missing region $\Omega_0$. Note that in every iteration only the missing pixels $y_p^u$ in the selected patch $y_p$ are inpainted and the missing region $\Omega_0$ is updated accordingly. The order in which the patches are selected for filling is based on the presence of structure and the number of known pixels. This helps in propagating the structure inside the missing regions as a result of which the global structure is preserved. One may also super-resolve all the patches in the known region by a factor of 2 by estimating the corresponding HR patches as explained in Section 6.2.2. This will result in an HR image where both known and inpainted regions are super-resolved.

## 6.3    EXPERIMENTAL RESULTS

We now present the inpainted results on the natural scene dataset available in Huang et al. [52]. The dataset also contains results of the state-of-the-art methods for image inpainting viz. image melding [23], Photoshop CS5 content aware fill [6], statistics of patch offsets [50], GIMP Resynthesizer plugin [48], planar structure guidance [51, 131] and the method by Komodakis and Tziritas [64]. We compare the results of our proposed method with these methods. The number of candidate matches considered in our implementation is $K = 5$ and the patch size is taken to be $m = 7$. The comparative results are presented in Figures 6.4–6.6 which are discussed below.

Figure 6.4 shows the results of inpainting the marked region corresponding to one of the kids in the cage. One can see the outline of the kid as well as the rods showing inconsistent bending in the inpainted results shown in Figures 6.4c–6.4d. An extra arm can be seen in Figure 6.4e while some artefacts can be seen in Figures 6.4f and 6.4i. The results in Figures 6.4g–6.4h are not only blurred, but also show inconsistency in the inpainted rods. The inpainted region in the proposed method shown in Figure 6.4j looks visually better when compared to other approaches.

Another result in Figure 6.5 shows the inpainting of benches on a hill-top. The result in Figure 6.5 shows unrealistic criss-cross shadows of the fence, while those in Figures 6.5c, 6.5f and 6.5h have shadows of the fence in the right-half of the image, which is undesirable. The result shown in Figure 6.5 is clearly not consistent with the known regions. Similarly, Figure 6.5e has the door extended downward that unrealistically cuts through the floor, while Figure 6.5i appears to have a visible seam on the boundary of the inpainted region. Note that the result of the proposed method in Figure 6.5j does not have any unrealistic shadows and is seamlessly inpainted. The texture of the inpainted region matches well with the region surrounding it.

In order to show the effectiveness of our approach in super-resolving in addition to inpainting we also present a result showing SR in Figure 6.6. The inpainted and super-resolved region is compared with Glasner et al.'s approach [42] where the SR is performed on our inpainted result at the original resolution. Note that SR approaches super-resolve only what is available, i.e., regions having no missing pixels, whereas the missing pixels are estimated and also super-resolved in our approach. Hence, our approach not only inpaints but also reconstructs high-resolution of the unknown region with missing pixels. We display the inpainted result in Figure 6.6d and simultaneous SR in Figure 6.6e obtained using our method. The expanded version after upsampling one

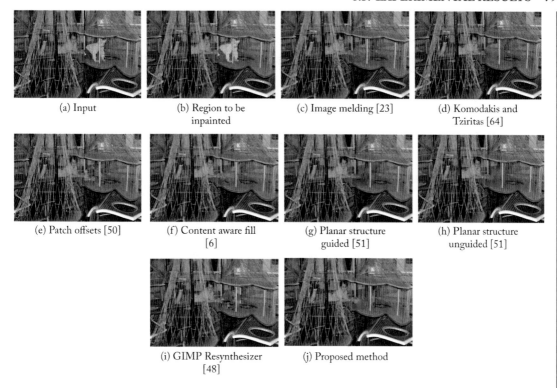

(a) Input

(b) Region to be inpainted

(c) Image melding [23]

(d) Komodakis and Tziritas [64]

(e) Patch offsets [50]

(f) Content aware fill [6]

(g) Planar structure guided [51]

(h) Planar structure unguided [51]

(i) GIMP Resynthesizer [48]

(j) Proposed method

**Figure 6.4:** Results of inpainting the marked region corresponding to one of the kids in cage.

of the inpainted regions (shown by the rectangular area with a blue boundary in Figure 6.6d) using bicubic interpolation and Glasner et al.'s method [42] for SR are depicted in Figures 6.6f and 6.6g, respectively. Looking at the results, we see that the super-resolved region shown in Figure 6.6h is comparable to the SR result shown in Figure 6.6g. Also, the simultaneously super-resolved region as obtained in our approach (Figure 6.6h) shows greater detail than simply upsampling the inpainted region using bicubic interpolation as shown in Figure 6.6f.

We now show the results of our method on heritage site images in Figures 6.7–6.8. Sword-marks over the historic stone carvings at the Sun temple in Modhera, India, are shown in Figure 6.7b. The inpainted image using our method is shown in Figure 6.7c. Note that the inpainting is seamless and our method generates plausible artistic work inside the inpainted regions. Also, the simultaneously inpainted and super-resolved expanded version (Figure 6.7d) of the rectangular area with yellow outline in Figure 6.7c provides clearer details in comparison to the bicubic interpolated expanded version (Figure 6.7d) of the inpainted region shown in Figure 6.7c. A damaged wall of a temple in Hampi, India, is marked as shown in Figure 6.7b. We display its inpainted version using our method in Figure 6.7c. Here one can observe the repair of a substantially large

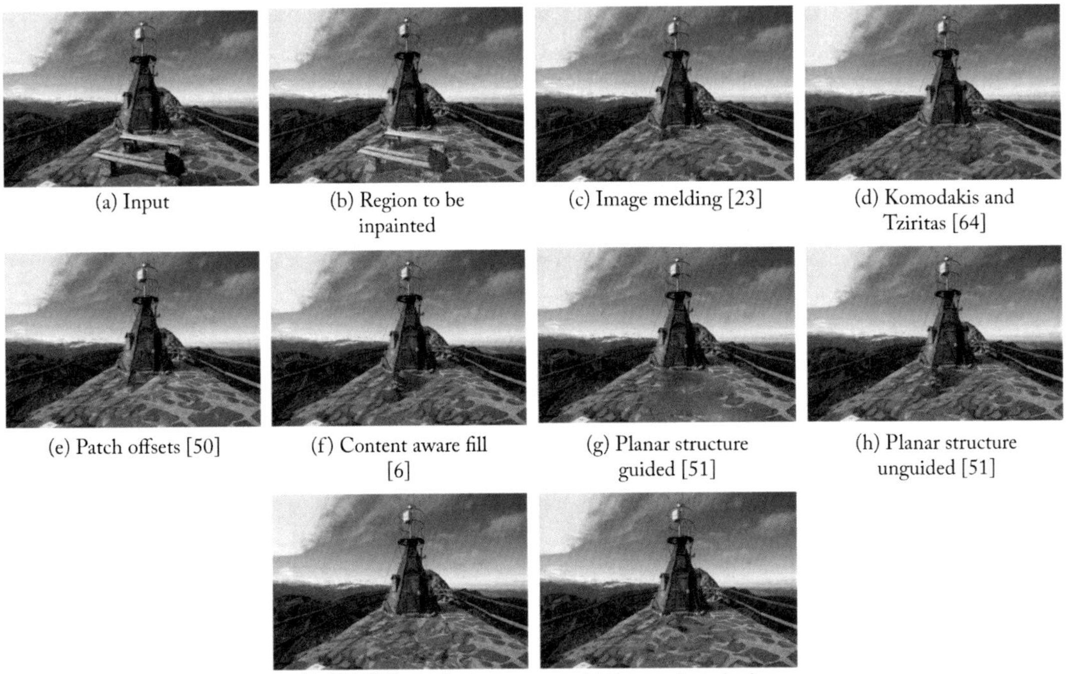

(a) Input

(b) Region to be inpainted

(c) Image melding [23]

(d) Komodakis and Tziritas [64]

(e) Patch offsets [50]

(f) Content aware fill [6]

(g) Planar structure guided [51]

(h) Planar structure unguided [51]

(i) GIMP Resynthesizer [48]

(j) Proposed method

Figure 6.5: Results of inpainting benches on the hill-top.

damaged region wherein the fine details within the repaired region are also visible. The expanded region shown in Figure 6.8e provides greater detail compared to the bicubic interpolated region shown in Figure 6.8d.

If the point cloud data corresponding to the input images in Figures 6.7a–6.8a were available, the damaged regions would be considered as holes and repaired using the Poisson surface reconstruction method [60]. Although this would fill the holes, it does not create artistic details inside large missing regions. However, if the missing regions are filled in the photographs, and higher resolution details are also made available as done above in our method, the resulting images can be used as sources for generating 3D models containing filled holes with reliable artistic details.

## 6.4   CONCLUSION

We have presented a unified approach to perform simultaneous inpainting and super-resolution. By using an additional constraint of matching patches at the original resolution as well as at the higher resolution, we not only obtain better source patches for inpainting but also have the cor-

**Figure 6.6:** Result showing simultaneous inpainting and SR: (a) input; (b) regions to be inpainted; (c) inpainting using planar structure guidance [51]; (d) inpainting using proposed method showing a rectangular area with blue boundary inside one of the inpainted regions; (e) simultaneously inpainted and super-resolved image (by a factor of 2) using the proposed method with known regions upsampled using bicubic interpolation; (f)–(h) expanded versions after upsampling (the region marked by the rectangular area with blue boundary in (d)) using various approaches viz. (f) bicubic interpolation, (g) Glasner et al.'s method [42] and (h) our method for super-resolution.

**Figure 6.7:** Result of simultaneous inpainting and super-resolution of the sword-marks at the Sun Temple in Modhera, India. (a) Input image; (b) regions to be inpainted are shown in red color; (c) inpainted result; (d) expanded version after bicubic interpolation of the region marked with a rectangular area with yellow boundary in (c); (d) expanded version of the simultaneously inpainted and super-resolved region (by a factor of 2) corresponding to the rectangular area with yellow boundary in (c).

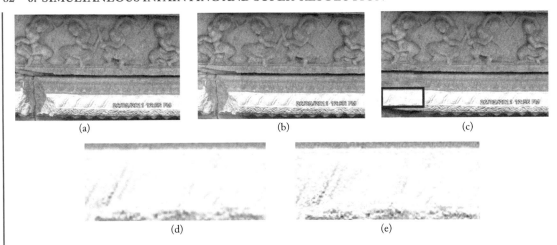

Figure 6.8: Result of simultaneous inpainting and super-resolution of a damaged wall of a temple in Hampi, India. (a) Input image; (b) regions to be inpainted are shown in red color; (c) inpainted result; (d) expanded version after bicubic interpolation of the rectangular area marked with blue boundary in (c); (e) expanded version of the simultaneously inpainted and super-resolved region (by a factor of 2) corresponding to the rectangular area marked with blue boundary in (c).

responding super-resolved version. A comparison with the state-of-the-art inpainting methods shows that the inpainted results of the proposed method are indeed better. Also, the simultaneously super-resolved regions are comparable to the super-resolution of the inpainted regions obtained using the method in Glasner et al. [42] and also show greater detail than those obtained by upsampling the inpainted regions using bicubic interpolation. The simultaneously inpainted and super-resolved images can be used as sources for generating 3D models with a higher amount of details.

CHAPTER 7

# Detecting and Inpainting Damaged Regions in Facial Images of Statues

In Chapters 4–6 we discussed inpainting techniques wherein the regions to be inpainted are manually provided by the users. When looking at heritage monuments, especially statues, there is a consensus about the desire to view these without any damage to the dominant facial regions. This encourages the exploration for a technique that automatically detects the damaged regions so that their repair can be completely automated using an existing inpainting technique. Another reason that motivates this exploration is that it can be a useful tool to continuously monitor the heritage site using a surveillance system and alert the authorities if any damage to monuments takes place. The damage could be either intentional or unintentional due to the curiosity of visitors, based on which a necessary action can be initiated and monuments can be protected from further damage.

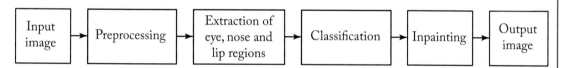

**Figure 7.1:** Block diagram of our approach for detecting and inpainting the damaged facial regions in statues.

In this chapter, we discuss a method that automates the process of detecting the damage to visually dominant regions viz. eyes, nose and lips in facial images of statues, and their repair using inpainting. Here the bilateral symmetry of the face is used as a cue to detect the eye, nose and lip regions. Textons features [124] are then extracted from each of these regions in a multi-resolution framework to characterize their textures. These textons are matched with those extracted from a training set consisting of true damaged and non-damaged regions in order to perform the classification. The repair of the identified damaged regions in the test image is then performed using the Poisson image editing method [97] by considering the best matching non-damaged region from the training set. A block diagram of our method is shown in Figure 7.1, the details of which are organized as follows. In Section 7.1 we discuss preprocessing, followed by extraction of the eye, nose and lip regions in Section 7.2. Details of classifying the detected region as either damaged or non-damaged are given in Section 7.3. We discuss the repair of the detected damaged regions using inpainting in Section 7.4 followed by a conclusion in Section 7.6.

## 7.1   PREPROCESSING

The input is assumed to be a frontal face image. One may note that slight deviation from the frontal pose is not an issue in the region detection process. However, for large deviation due to complex distortions, one may think of using image registration as a preprocessing step. Nevertheless, given a single image with complex distortions, registration itself is a difficult problem and involves pixel interpolation, affecting classification based on texture. We therefore consider only the frontal face images as our inputs.

The detection of the regions of interest viz. eyes, nose and lips is based on edge features and can get affected by changes in the illumination conditions. In order to make the detection process robust to illumination changes, we apply the single scale retinex (SSR) algorithm[1] [57] on the input image. Further, we apply an edge-preserving smoothing operation [98] over the resulting image, so as to detect the regions with better accuracy. Following this, the edges are extracted to obtain an edge image $I_e$.

## 7.2   EXTRACTION OF EYE, NOSE AND LIP REGIONS

The visually dominant regions, viz. eyes, nose and lips, have a common property of being bilaterally symmetrical. Motived by the work in Katahara and Aoki [59], our approach uses this property as a cue for detecting the eye, nose and lip regions. Using the edge image $I_e$, symmetry measures $b_h(x, y)$ and $b_v(x, y)$ around each pixel location $(x, y)$ are calculated in the horizontal and vertical directions, respectively as follows,

$$
\begin{aligned}
b_h(x, y) &= \sum_{j=1}^{\min(y, y-N)} [1(I_e(x, y-j) = I_e(x, y+j))] \text{ and} \\
b_v(x, y) &= \sum_{i=1}^{\min(x, x-M)} [1(I_e(x-i, y) = I_e(x+i, y))],
\end{aligned}
\tag{7.1}
$$

where, $M \times N$ represents the size of input image and $1(condition)$ is an indicator function that outputs the value of 1 if $condition$ is true, else outputs 0.

The calculated symmetry measures are then used to obtain the projections $S_x$ and $S_y$ as follows:

$$
S_x(y) = \sum_{i=1}^{M} b_h(i, y), \quad \text{and} \quad S_y(x) = \sum_{j=1}^{N} b_v(x, j),
\tag{7.2}
$$

---

[1]We used the implementation of SSR algorithm available at: http://in.mathworks.com/matlabcentral/fileexchang e/26523-the-inface-toolbox-v2-0-for-illumination-invariant-face-recognition/content/INface_too l/photometric/single_scale_retinex.m [125].

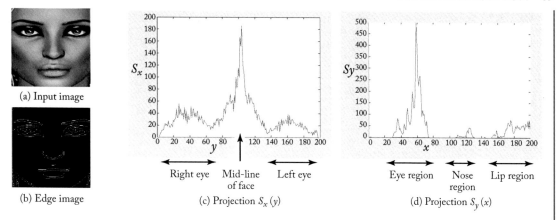

(a) Input image

(b) Edge image

(c) Projection $S_x(y)$

(d) Projection $S_y(x)$

**Figure 7.2:** Extraction of the potential regions of interest using bilateral symmetry.

where, $y$ and $x$ respectively denote the column and row being projected. The peak in projection $S_x$ provides the mid-line about which the face is nearly symmetric, while the peaks in projection $S_y$ help in identifying vertical locations of the eye, nose and lip regions. This is illustrated using the example shown in Figure 7.2. The regions of interest can then be extracted using appropriately sized windows around the locations of the peaks detected in projections $S_x$ and $S_y$.

## 7.3    CLASSIFICATION

For classifying the detected regions as damaged or non-damaged we use texture as a cue. A method for modeling different texture classes that have uniformity within each class has been proposed in Varma and Zisserman [124]. Our work, however, deals with the images of statues at historic monuments that have natural textures with no uniformity. In such cases, it is difficult to extract any repetitive pattern at a single scale. However, irregular patterns and structures in nature have been successfully represented using fractals [29, 69]. The fractals are geometric patterns that repeat at smaller scales to produce irregular shapes and surfaces that cannot be represented by classical geometry. This motivated us to make use of a multi-resolution framework to address the issue of irregularities in natural texture at different resolutions—a property characterized by stone-work and monument surfaces. Moreover, our method automatically calculates the number of clusters required to represent the two classes that correspond to damaged and non-damaged regions, as opposed to the approach in Varma and Zisserman [124] which uses a fixed number of clusters for representing several texture classes.

The texture features are extracted in the form of textons, which are cluster centers in the filter response space. These textons are obtained in a multi-resolution framework by convolving the detected potential region of interest and its two coarser resolution versions with the maximum-response-8 (MR8) filter bank [124]. In order to obtain the coarser versions of the detected region,

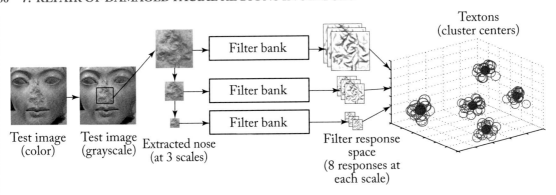

**Figure 7.3:** Texton extraction from the detected nose region in a test image.

it is low-pass filtered using a Gaussian filter before downsampling. The MR8 filter bank consists of 38 filters viz. edge and bar filters with 6 orientations at 3 scales along with a Gaussian and a Laplacian of Gaussian filter. Each pixel of the input region is now transformed into a vector of size 8 by considering 8 maximum responses out of the 38 filters. In other words, the maximum response for orientation of the edge and bar filters at each scale along with the response for the Gaussian and Laplacian of Gaussian filters are recorded to obtain a vector of size 8. We illustrate the process of extracting the textons for a detected nose region in the test image, with the help of Figure 7.3. A similar process is independently applied to extract the texton features from the eye and lip regions.

The K-means algorithm is then applied on these vectors to obtain the $K$ cluster centers, i.e., textons. One may note that the method proposed in Varma and Zisserman [124] requires the number of clusters ($K$) to be known in advance. However, it may not be possible to pre-determine the number of clusters $K$ as this is a data-dependent term. In our work, we use a simple approach to estimate the optimal number of clusters. Here we plot a two-dimensional evaluation graph, where the X-axis shows number of clusters ($K$) and the Y-axis shows the pooled within cluster sum of squares around the cluster means ($W_k$) calculated as follows [118]:

$$W_K = \sum_{r=1}^{K} \left( \sum_{\forall i,i' \in C_r} d_{i,i'} \right), \tag{7.3}$$

where $d_{i,i'}$ is the squared Euclidean distance between members $(i, i')$ of cluster $C_r$. Here Tibshirani et al. [118] have shown that the point at which the monotonic decrease flattens markedly provides the optimal value of $K$. However, if the curve is smooth, it is difficult to determine where exactly this decrease flattens. We then have a challenging task to obtain the optimal value of $K$. To overcome this difficulty, we attempted to best fit two straight lines to the curve using expectation-maximization (EM) algorithm. The point of intersection of the two best fit lines then

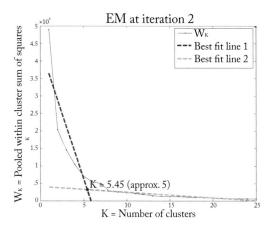

**Figure 7.4:** Auto-selection of number of clusters $K$ by fitting two straight lines to the data.

gives the approximate point at which the curve starts to flatten. The projected point on the axis of number of clusters is then considered as the optimal value of $K$ as illustrated in Figure 7.4.

A process as described above is used offline for extracting the textons from a training set consisting of true damaged and non-damaged regions. Here the textons representing a damaged eye, nose or lip region are extracted using all the training images containing the corresponding true damaged region. Likewise, textons representing the non-damaged regions are extracted using the true non-damaged regions from all the training images. An example showing the offline extraction textons from a training set containing damaged nose regions is illustrated in Figure 7.5. We now compute the Euclidean distance between textons of the detected region (viz. eye, nose or lip) in the test image and those from the corresponding true damaged and non-damaged regions of training images, to perform the classification. Here the minimum distance criteria is used to classify the region as either damaged or non-damaged. It may be noted that for each extracted region, viz. eyes, nose and lips, the classification is performed independently. This enables the simultaneous detection of multiple damaged regions in the test image.

## 7.4 INPAINTING

Once a region is identified to be damaged, it needs to be inpainted using a non-damaged source region. Here if one eye is damaged we use the flipped version of the other eye (detected automatically) from the same image as the source. However, if both eyes or the nose or lip regions are damaged, we make use of the images from the training set as the source for inpainting. Here the source selection criteria is the extent of similarity in the Euclidean space, of the undamaged region in the image containing the detected damaged region, with the true undamaged regions in the training set images. However, if all the detected regions in an image are damaged, then the

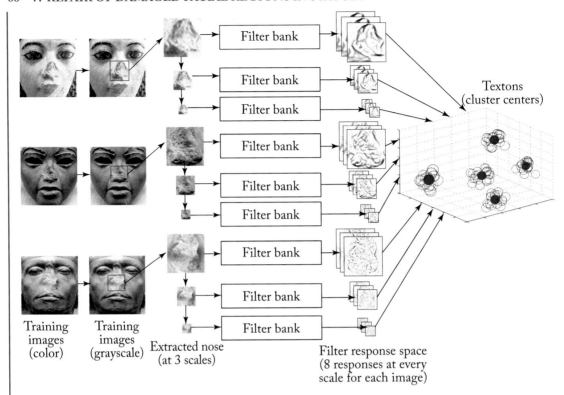

**Figure 7.5:** Offline extraction of textons from the training set images containing damaged nose regions.

source regions need to be provided manually. Once we have the source region, we use it as a guidance vector field for the Poisson image editing technique [97] to inpaint the identified damaged region.

## 7.5   EXPERIMENTAL RESULTS

We now discuss the results of our experiments conducted on a database consisting of 40 facial images of Egyptian statues having damaged and non-damaged regions, downloaded from the Internet [44]. The spatial resolution of the images is adjusted such that all images are of the same size. A mean correction is applied to the images so that they have the same average brightness. Training for the eye, nose and lip regions was done independently. For training we have used 10 images each for damaged and non-damaged regions. Testing was carried out on all the images from the database including those used for training.

The results using our approach are shown in Figures 7.6–7.9. The detection and inpainting of a damaged nose is shown in Figure 7.6, where the source used for inpainting is an image from the training set containing an undamaged nose. In Figure 7.7, the reflected version of the non-damaged left eye has been used to inpaint the damaged right eye. However, in Figure 7.8 since both eyes are damaged, an image from the training set containing non-damaged eyes is used as the source for inpainting. Note that the criteria used for selecting the source is similarity of the non-damaged regions in the test image with the corresponding regions in the images from the training set. In Figure 7.9, we show a result where our method fails to detect the damaged nose. Here the input image contains the nose region having a small amount of damage, due to which the corresponding textons match those of the non-damaged nose regions from the training set. This is caused by the extracted statistics of the damaged and non-damaged regions. Thus, among the extracted potential regions of interest shown in Figure 7.9b, the damaged nose is incorrectly classified as undamaged and is therefore undetected in Figure 7.9c.

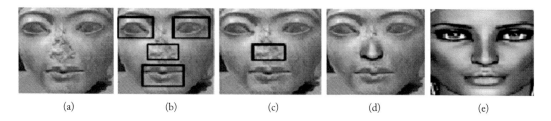

(a)              (b)              (c)              (d)              (e)

**Figure 7.6:** Detecting and inpainting a damaged nose; (a) input image, (b) extracted potential regions of interest, (c) detected damaged nose, (d) inpainted nose using the source image (e).

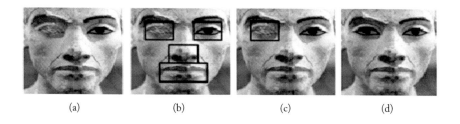

(a)              (b)              (c)              (d)

**Figure 7.7:** Detecting and inpainting a damaged eye; (a) input image, (b) extracted potential regions of interest, (c) detected damaged eye, (d) inpainted eye.

We now discuss the performance of our method of automatic detection of facial regions and inpainting by considering the ground truth from the inputs provided by the volunteers. Performance evaluation is done in terms of the standard recall and precision metrics defined as: Recall$= \frac{|Ref \cap Dect|}{|Ref|}$ and Precision$= \frac{|Ref \cap Dect|}{|Dect|}$. Here $Ref$ are the regions declared to be damaged or undamaged by volunteers and $Dect$ are the regions detected as damaged or undamaged by the

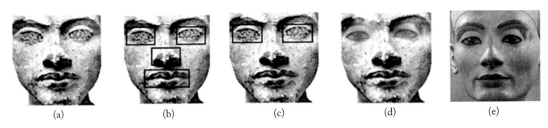

**Figure 7.8:** Detecting and inpainting damaged eyes; (a) input image, (b) extracted potential regions of interest, (c) detected damaged eyes, (d) inpainted eyes using the source image (e).

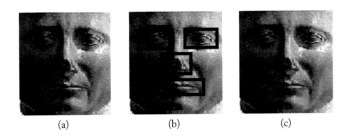

**Figure 7.9:** Failure case; (a) input image, (b) extracted potential regions of interest, (c) damaged nose is incorrectly classified as undamaged.

proposed technique. From a set of 40 images, 50 regions were found to be damaged, while 50 were undamaged. Out of 50 damaged regions 47 were correctly classified, while all 50 undamaged regions were correctly classified. For source region selection, 49 out of 50 regions were correctly selected. The performance in terms of the recall and precision metrics is summarized in Table 7.1 which shows the effectiveness of our method.

**Table 7.1:** Performance in terms of recall and precision

| Region Type | # Regions | Recall | Precision |
|-------------|-----------|--------|-----------|
| Damaged | 50 | 0.9400 | 1.0000 |
| Undamaged | 50 | 1.0000 | 1.0000 |

Note that the source selection method used in our approach is not comparable with content-based image retrieval (CBIR) techniques. This is because, for a large damaged region, a CBIR system may not find an adequate amount of non-damaged content to retrieve a good match relevant for inpainting. Although our method is developed for images of statues, it can be effective for facial regions in natural images as both have the same facial characteristics. Thus, we have presented a texture-based approach to automatically detect the damaged regions in facial images of statues for performing their digital repair using an existing inpainting technique. The results show that these regions can be effectively repaired.

## 7.6   CONCLUSION

In this chapter, we have presented a technique for automatic detection of damaged dominant facial regions in statues for their digital repair. Here a bilateral symmetry-based method is used to identify the eyes, nose and lips and the texton features are extracted from each of these regions in a multi-resolution framework to characterize the textures of damaged and non-damaged regions. These textons are matched with those extracted from a training set of true damaged and non-damaged regions for detecting the damaged ones which are then inpainted with the help of suitable source regions. Here we have addressed the repair of specific regions viz. the facial regions of statues in heritage monuments. However, damage like cracks in the non-facial regions of the monuments also diminishes their attractivenesses. We address this in the following Chapter 8 wherein we present techniques for automating the digital repair of cracks.

# CHAPTER 8

# Auto-inpainting Cracks in Heritage Scenes

In Chapter 7 we discussed a technique for automatic detection and inpainting of the damaged eye, nose and lip regions in facial images of statues. In this chapter, we consider the damage in non-facial regions and describe techniques to automatically detect these and inpaint them. In particular, we consider the non-facial region that can have damage like cracked regions. These cracks in a heritage site could be developed over a period of time due to environmental effects or due to manual destruction. Such damage can also diminish the attractiveness of heritage monuments and one would want the cracks to be seamlessly eliminated. Automatic detection of cracks in the acquired images will be particularly useful for performing on-the-fly digital reconstruction when the tourists take the pictures of the heritage monuments using their handheld image/video-capturing devices. Toward this end, in this chapter we discuss techniques for automatic detection of cracked regions and demonstrate their repair using existing inpainting algorithms [21, 88]. Note that the application is to actually restore a heritage scene, i.e., digitally repair cracks that physical objects have. Thus, we are not talking about image restoration, but about object completion. In other words, we do not detect an external damage or defect due to alteration of a photograph, but instead detect and inpaint the cracked regions in the photographed scenes/objects.

The contents of this chapter are organized as follows. In Section 8.1 we discuss a simple yet effective method to automatically detect the damaged regions which are characterized by abruptly dark deteriorations in the photographed monuments of a heritage site. The use of order-statistics and density filters makes our technique computationally fast and we are able to accurately detect the cracked regions. In Section 8.2 we discuss a crack detection method that is based on comparing overlapping adjacent patches using singular value decomposition (SVD). Another effective and more accurate crack detection technique is discussed in Section 8.3. This technique is based on patch comparison using a measure derived from the edit distance, which is a popular string metric used in the area of text mining. In Section 8.4 we extend our crack detection approach to perform inpainting of video frames by making use of the scale invariant feature transform (SIFT) and homography. We consider the camera movement to be unconstrained while capturing video of the heritage site, as such videos are typically captured by novices, hobbyists and tourists. Here we also provide a video quality measure to quantify the quality of the inpainted video.

## 8.1 A SIMPLE METHOD FOR DETECTING AND INPAINTING CRACKS

Images of heritage sites usually contain murals and monuments that have a porous surface, making the detection of the damaged region exhibiting visual discontinuities a very challenging task. Here we discuss a computationally efficient and effective technique to detect these visual discontinuities which appear as abruptly dark regions representing the cracks. A block diagram of our method for crack detection is shown in Figure 8.1, the details of which are discussed below.

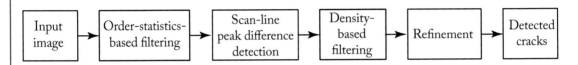

Figure 8.1: A simple approach for crack detection.

Given an input image $I$ of a damaged monument in the RGB color space, we transform it to the *HSV* color space and extract the grayscale image $I_V$ that corresponds to the *intensity* image. In the images of heritage monuments, the damaged regions like cracks usually appear darker in comparison to their surrounding areas. However, due to the porous surface of photographed monuments, one observes large variations in intensity of many adjacent pixels. It is therefore necessary to enhance the contrast of these dark regions with respect to the surrounding regions. A widely used method for contrast enhancement is the histogram equalization. This is followed by edge detection to detect intensity contrast among neighboring pixels [43]. Our experiments, however, revealed that contrast enhancement using histogram equalization increased the contrast of various other regions as well, and subsequent edge detection resulted in many false positives.

### 8.1.1 ORDER-STATISTICS-BASED FILTERING

Order-statistics filters have been used for edge enhancement in noisy images [67, 68]. This motivated us to use order-statistics filters instead of histogram equalization for enhancing the contrast of dark regions. There are various filters based on order-statistics, such as the min, max and *median* filters. We use a combination of the min and max filters to achieve the desired contrast enhancement. Since the min and max filters enhance the dark and bright regions in the image, by taking an average of the output of these filters we obtain a smoothed version $I_a$ of the intensity image $I_V$. This helps us to overcome the rapid variation in intensity due to porous regions. The max filtered version of $I_a$ is then subtracted from the intensity image to achieve the contrast enhancement of the abruptly dark regions. We achieve this by considering a patch $\Phi_p$ around every pixel $p$ in the image $I_V$ and obtain the min and max filtered images by extracting the minimum and maximum intensities from each patch $\Phi_p$. The average of these filters, i.e., $I_a$, is obtained using

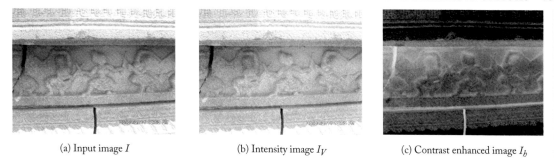

(a) Input image $I$                (b) Intensity image $I_V$                (c) Contrast enhanced image $I_h$

**Figure 8.2:** Contrast enhancement using order-statistics-based filtering.

the Equation (8.1) as follows.

$$I_a(p) = \frac{\min(\Phi_p) + \max(\Phi_p)}{2}, \quad \Phi_p \in I_V. \tag{8.1}$$

The contrast-enhanced image $I_h$ is then obtained as follows using Equation (8.2),

$$I_h(p) = I_V(p) - \max(\Phi_p), \quad \Phi_p \in I_a. \tag{8.2}$$

The input image $I$ and the contrast-enhanced image $I_h$ are shown in Figures 8.2a and 8.2c, respectively. One may note that the contrast-enhanced image $I_h$ is not the same as the intensity image with inverted thresholds. Due to the porous surface of the photographed monuments, the intensity image $I_V$ (shown in Figure 8.2b) and its thresholds-inverted version contain random sharp variations in contrast. Whereas, the image obtained after contrast enhancement as shown in Figure 8.2c exhibits sharp contrast variations at the regions with abrupt changes in the intensity, with smooth contrast variations elsewhere. Thus, we observe a drastic variation in contrast of the cracks with respect to their surrounding regions.

## 8.1.2    SCAN-LINE PEAK DIFFERENCE DETECTION

Since the contrast-enhanced image $I_h$ exhibits sharp variations at the potentially damaged regions, one can detect these by comparing adjacent pixel values along a scan-line. Large peaks of such a comparison between adjacent pixel values of a scan-line correspond to the potential cracks on that scan-line. In order to detect these cracks, we consider every scan-line to contain at most $\alpha$ number of peaks to be a part of the cracked regions. Here we use a set $M_x$ for denoting these $\alpha$ number of peaks per scan-line as follows,

$$\begin{aligned} M_x = \quad &\{\lambda_j, j = 1, \ldots, \alpha | \lambda_1 \geq \lambda_2 \geq \ldots \geq \lambda_n, \\ &\lambda_k = d_k(x, y), k = 1, \ldots, n\}, \quad \text{where,} \end{aligned} \tag{8.3}$$

$$d(x, y) = \quad \max(|I_h(x, y) - I_h(x, y + i)|), i = \pm 1,$$

where $\lambda_1, \lambda_2, \ldots, \lambda_n$ are the pixel value differences between the adjacent columns $y$ at every scan-line $x$, arranged in descending order. The parameter $\alpha$ controls the number of high peak-differences, i.e., pixels in the potential cracks, that can be detected along a scan-line. Using the set $M_x$ per scan-line we create a binary image $I_b$ denoting the candidate pixels in the potential cracks as,

$$I_b(x, y) = \begin{cases} 1, & \text{if} \quad d(x, y) \in M_x \\ 0, & \text{otherwise.} \end{cases} \tag{8.4}$$

### 8.1.3  DENSITY-BASED FILTERING

Using such a process for peak-difference detection along a scan-line detects pixels at nearby locations in adjacent scan-lines. This leads to an unwanted increase in the density of the detected pixels around the potential cracks. We now use a density-based filter to threshold a local region $\varphi_p$ around every pixel $p$ detected in the binary image $I_b$ based on the presence of other detected pixels inside $\varphi_p$. This discards the isolated pixels detected in $I_b$ which do not belong to the potential cracks. Thus we localize the cracks by creating another binary image $I_d$ as follows,

$$I_d(\varphi_p) = \begin{cases} 1, & \text{if} \quad \sum_{q \in \varphi_p} \frac{I_b(q)}{|\varphi_p|} \geq \theta_1 \\ 0, & \text{otherwise,} \end{cases} \tag{8.5}$$

where $|\varphi_p|$ is the area of the local region $\varphi_p$ around a pixel $p$ and $\theta_1$ is the threshold density. The binary image $I_b$ and the density-filtered binary image $I_d$ are shown in Figure 8.3.

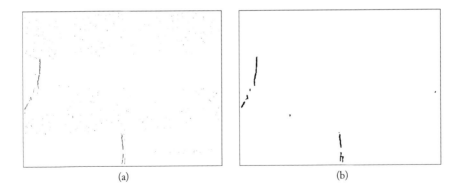

(a)        (b)

**Figure 8.3:** Detection of candidate pixels in the potential cracked regions. (a) The binary image $I_b$ generated using Equation (8.4) by detecting peak differences along every scan-line in the contrast enhanced image $I_h$; (b) the density-filtered binary image $I_d$.

### 8.1.4  REFINEMENT

The density-filtered binary image $I_d$, however, consists of disjoint parts of the cracks. In order to address this issue, we use the morphological dilation process so that the disjoint regions get connected. By doing so, few isolated group of pixels that do not belong to the cracked regions may also get connected. These are eliminated by applying yet another density-based filtering operation on the centroids of each connected components in $I_d$ to get a final binary image $I_g$ containing the detected cracks. To do this, let us denote the dilated version of $I_d$ by $I_D$ and a patch around centroid $C_i$ of every connected region $i \in I_D$ by $\widehat{\varphi}_i$. The density-based filtering operation on $I_D$ is then performed using Equation (8.6) as follows,

$$I_g(\widehat{\varphi}_i) = \begin{cases} I_D(\widehat{\varphi}_i), & \text{if} \quad \sum_{q \in \widehat{\varphi}_i} \frac{I_D(q)}{|\widehat{\varphi}_i|} \geq \theta_2 \\ 0, & \text{otherwise.} \end{cases} \tag{8.6}$$

Here $I_g$ is the output of the density-filtering operation on $I_D$, $|\widehat{\varphi}_i|$ is the area of the patch $\widehat{\varphi}_i$ and $\theta_2$ is the threshold.

(a) Input image $I$ overlaid with the crack detected in $I_g$ shown in red color

(b) Inpainted version of $I$ using the region detected in $I_g$

**Figure 8.4:** Automatic detection and repair of cracks.

The regions detected in $I_g$ represent the cracks and can therefore be used as a mask for inpainting. One can now use a suitable inpainting algorithm to perform their digital repair. We demonstrate the repair of the detected cracks using the inpainting [88] technique discussed in Chapter 4. Figure 8.4 illustrates the regions detected in the binary image $I_g$ overlaid on the input image $I$ and its inpainted version using the detected regions and the inpainting technique discussed in Chapter 4. One may note that the objective of this method is to detect abruptly dark regions that characterize cracks and is therefore not suited for detection of vandalized eye, nose and lip regions of faces in statues as discussed in Chapter 7, for which texture is used as a cue. Hence, the technique discussed in Chapter 7 is not suited for addressing the detection of cracks.

## 8.1.5    EXPERIMENTAL RESULTS

We present the results of our technique on the data collected from the world heritage site in Hampi, India. The images were captured using a Samsung ES55 digital camera. The data consists of a number images of monuments that have both cracked and non-cracked regions. Fairly large cracks are visible in all the images. The images used in the experiments are of size $684 \times 912$ pixels. For the order-statistics-based filtering operation we considered patches $\Phi$ of size $3 \times 3$. For detecting the peak difference in the scan-line of the contrast-enhanced image $I_V$, we have set the value of $\alpha$ to 2. The patch size $\varphi$ for the density-based filtering operation is taken to be $9 \times 9$ while for refinement $\widehat{\varphi}$ is chosen as $61 \times 61$. The thresholds $\theta_1$ and $\theta_2$ are set to 0.08 and 0.45, respectively.

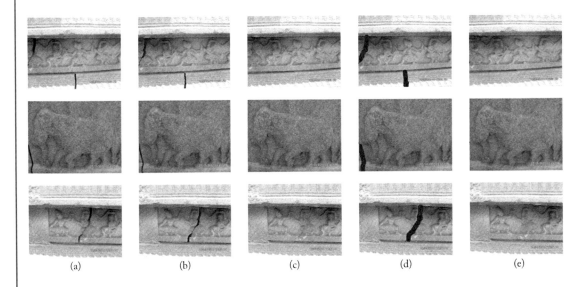

Figure 8.5: Results: (a) input images; (b) cracks marked by the volunteers are shown in red color; (c) inpainted version of (b); (d) automatically detected cracks using our method are shown in red color; (e) inpainted version of (d).

Since no ground truth data is available that presents the true pixels in the cracked region, we have manually determined them with the help of volunteers. Although the subjectively generated ground truth data may not be accurate, it can still be used to judge the effectiveness of the detected cracks. The results of this approach in comparison to the manually marked cracks by volunteers are shown in Figure 8.5. Here Figure 8.5a shows the input images. The cracks marked manually by volunteers are shown in Figure 8.5b, while Figure 8.5c shows the corresponding inpainted images. The cracks detected automatically using our technique are shown in Figure 8.5d, while Figure 8.5e shows their inpainted versions. The results reported in Figure 8.5 show that automatically detected cracked regions cover almost all the pixels marked as cracks by the volunteers. Use of these regions as input mask for inpainting techniques is justified from the inpainted results.

We now quantify the accuracy of the detected cracks using the standard recall and precision metrics defined in Equation (8.7) as follows,

$$Recall = \frac{|Ref \cap Dect|}{|Ref|} \quad \text{and} \quad Precision = \frac{|Ref \cap Dect|}{|Dect|}, \tag{8.7}$$

where *Ref* are the pixels declared to be in the cracked regions by volunteers and *Dect* are the pixels detected as cracks by our technique. For detected cracks in the binary image $I_g$ to be suitable for use in an inpainting algorithm, the desired *Recall* value should be nearer to 1. A lower value indicates that fewer pixels in the cracked regions have been detected and therefore the undetected cracked pixels are used as a source for inpainting, leading to poor inpainted results. On the other hand, a low *Precision* value can be acceptable as it indicates that more number of pixels have been detected, which only increases the size of the region to be inpainted. The *Recall* and *Precision* values for the results shown in Figure 8.5d along with the time taken for detection are given in Table 8.1. Here we observe that the detected cracks have a high *Recall* value, indicating that almost all the cracked pixels marked by the volunteers are detected.

Table 8.1: Performance evaluation for the automatically detected cracks shown in Figure 8.5d

| Images | # Cracked Pixels Marked by Volunteers | Recall | Precision | Time (sec) |
|---|---|---|---|---|
| Image in row 1 | 6,501 | 0.9812 | 0.3254 | 3.3852 |
| Image in row 2 | 3,529 | 0.9717 | 0.2986 | 3.4164 |
| Image in row 3 | 5,500 | 0.9955 | 0.3082 | 3.3440 |

We now compare our results with those obtained from the defect-detection method proposed in Amano [3] as shown Figure 8.6. It may be noted that the results for the technique proposed in Amano [3] are obtained after fine-tuning the parameters. Yet we observe that the defect-detection method in Amano [3] detects superimposed text but fails to detect the cracks. On the other hand, the regions detected by our method are similar to the regions marked as cracks by the volunteers. This is also suggested by the *Recall* and *Precision* values given in Table 8.2. Moreover, the timing information suggests that the detection of cracks using our method is significantly fast. Here both the techniques have been implemented on the same machine with an Intel Core i5 processor.

We now present the detection results for images that do not contain the cracked regions in Figure 8.7. Here we again observe that the defect-detection technique in Amano [3] detects the superimposed text. On the other hand, our method does not detect any region, thus avoiding false detection. In other words, our method is also capable of correctly identifying cases in which no cracks are present.

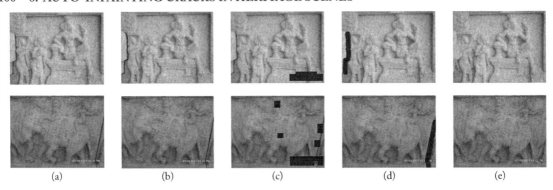

    (a)          (b)          (c)          (d)          (e)

**Figure 8.6:** Comparative results with respect to the technique proposed in Amano [3]. (a) Input images; (b) regions marked as cracks by volunteers; (c) detection using the technique in Amano [3]; (d) detection using our method; (e) inpainted version of (d).

**Table 8.2:** Performance comparison with respect to the method in Amano [3] for the results shown in Figure 8.6

| Images | #Target Pixels | Defect-detection Method [3] | | | Our Method | | |
|---|---|---|---|---|---|---|---|
| | | Recall | Precision | Time (sec) | Recall | Precision | Time (sec) |
| Image in row 1 | 3,494 | 0.0000 | 0.0000 | 109.00 | 0.9825 | 0.1763 | 3.5256 |
| Image in row 2 | 2,831 | 0.3727 | 0.0239 | 096.00 | 0.9414 | 0.1284 | 3.3384 |

We now analyze the computational complexity of our proposed technique. Consider the grayscale intensity image $I_V$ to be of size $M \times N$, having $\eta = MN$ number of pixels, while the patch $\Phi$ has $\kappa$ number of pixels such that $\kappa << \eta$. The number of pixels processed by each of the order-statistic filters viz. min and max is $\eta\kappa$. Also $2\eta\kappa$ number of pixels are processed when calculating $I_a$ which is the average of outputs of the order-statistic filters. Likewise, the calculation of the contrast-enhanced image $I_h$ involves the processing of $2\eta\kappa$ number of pixels. Further, the generation of the binary map $I_b$ requires the processing of $2\eta$ number of pixels. Let $k$ denote the total count of pixels processed in all further operations, which is the product of number of pixels detected in the binary map $I_b$ and the size of neighborhood for each operation. However, both these quantities are substantially small in comparison to $\eta$ and therefore their product $k << \eta$. The total number of pixels processed by our technique is therefore $T = 5\eta\kappa + 2\eta + k$. If processing each pixel takes a unit of time, then total time required by our technique is $5\eta\kappa + 2\eta + k = O(\eta)$, which is linearly proportional to the size of the grayscale intensity image $I_V$.

From the results shown in Figures 8.5–8.7, along with the performance comparison shown in Tables 8.1 and 8.2, it is clear that the cracked regions are successfully detected by our method. Although the technique in Amano [3] is good for detection for an alteration to the photograph

(a)                              (b)                              (c)

**Figure 8.7:** Results for images that do not contain cracks. (a) Input images; (b) detection result of the technique proposed in Amano [3]; (c) detection results using our technique.

(like overlay text), our method is more suitable when it comes to detection of damage in the photographed scene/object and is comparatively faster. For successful commercialization of the auto-detection of cracks, one needs to obtain the results in quick time. Our method is straightforward and the speed of our approach is linearly proportional to the size of the input image, making it usable in real-time applications. This makes our algorithm suitable for use in digital cameras to obtain on-the-fly automatic detection of the cracks and their inpainting when a tourist is capturing the photograph of a damaged heritage scene.

## 8.2 SINGULAR VALUE DECOMPOSITION-BASED CRACK DETECTION AND INPAINTING

In the previous Section 8.1 we discussed a simple yet effective technique for detecting and inpainting the cracked regions in heritage monuments. The method is intuitive and relies on the detection of drastic variations in the intensity of adjacent pixels across a scan-line to detect the boundary of the cracked regions. Yet, it is a primitive approach that performs pixel-based comparisons and involves many parameters that are set heuristically. In this section, we discuss a patch-based method for detecting the cracks. Here the main idea is to compare the overlapping patches for similarity using the singular value decomposition (SVD)-based patch analysis. Calculating an average dissimilarity of the row and column adjacent patches with respect to a patch under consideration helps to reveal the amount of visual discontinuity between the patches. By using a threshold, the cracks can then be identified as the ones having higher dissimilarity value. Once the cracks are detected, their filling is demonstrated using an existing inpainting technique [21]. Figure 8.8 shows our SVD-based technique to automatically detect the cracks for inpainting, the details of which are discussed below.

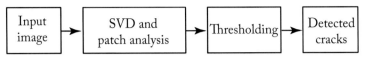

Figure 8.8: SVD-based approach for crack detection.

Given an input image $I$ in the RGB color space, we first transform it into *HSV* color space and extract the grayscale image $I_V$ that corresponds to the *intensity* image. Now consider a patch $\Phi_p$ of size $m \times n$ at pixel $p \in I_V$ with coordinates $(x, y)$. Here $x = 1, \ldots, M - m$ and $y = 1, \ldots, N - n$, such that $M \times N$ represents the size of the image $I_V$. The elements of patch $\Phi_p$ are rearranged to form a column vector $v_p$ of length $L = mn$ by using lexicographical ordering of pixels. Similarly, consider the pixels $r$ and $s$ adjacent to $p$ with respective coordinates as $(x, y + 1)$ and $(x + 1, y)$. The corresponding patches $\Phi_r$ and $\Phi_s$ are also ordered lexicographically to obtain the vectors $v_r$ and $v_s$ as illustrated in Figure 8.9.

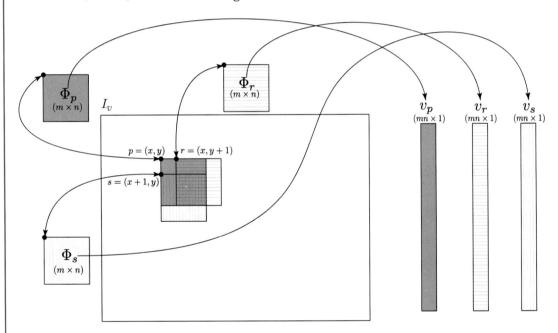

Figure 8.9: Overlapping patches considered for comparison.

We find the similarity between the vectors $v_p$, $v_r$ and $v_s$ using the geometric interpretation of the SVD model [24] on the matrix that has these vectors as its columns. By calculating the similarity between vectors of adjacent patches, we create a similarity matrix $S$ whose elements are then compared with an automatically estimated threshold $\delta$, to detect patches having discontinuities. In the following subsection we discuss patch analysis in the SVD domain.

## 8.2.1    SVD AND PATCH ANALYSIS

We form a matrix $A$ with the columns as $v_p$, $v_r$ and $v_s$ corresponding to patches $\Phi_p$, $\Phi_r$ and $\Phi_s$, and decompose it using SVD such that $A = U \Sigma V^T$. Here $U$ is an $L \times L$ matrix, the columns of which are the eigenvectors of $AA^T$, $V$ is a $3 \times 3$ matrix consisting of eigenvectors of $A^T A$, and $\Sigma$ is $L \times 3$ matrix of singular values ($\sigma_1 \geq \sigma_2 \geq \sigma_3 \geq 0$) at diagonals. We now reduce the size of matrices $U$ to $L \times 3$ and $\Sigma$ to $3 \times 3$, which however does not affect the reconstruction of $A = U \Sigma V^T$. By discarding the smallest eigenvalue, we further reduce the size of matrices $U$ to $L \times 2$, $\Sigma$ to $2 \times 2$ and $V$ to $3 \times 2$, which now leads to an approximate reconstruction of matrix $A$. Such a method is widely used for image compression and noise reduction [127].

Considering the updated matrices $\Sigma$ and $V$, the rows $w_1$, $w_2$ and $w_3$ of the matrix $V \Sigma$ now reflect the extent to which pixels in the corresponding columns $v_p$, $v_r$ and $v_s$ have a similar pattern of occurrence [24]. The extent of similarity between any two columns of the matrix $A$ is therefore given by the cosine of angle between corresponding rows of the matrix $V \Sigma$ as follows,

$$\cos(\theta_{pr}) = \frac{w_1 . w_2}{||w_1|| \, ||w_2||} \quad \text{and} \quad \cos(\theta_{ps}) = \frac{w_1 . w_3}{||w_1|| \, ||w_3||}. \tag{8.8}$$

Note that if the smallest eigenvalue is retained then the complete reconstruction of $A$ is possible, and in that case the angle obtained by directly considering the columns of matrix $A$ is the same as that obtained between the corresponding rows of matrix $V \Sigma$. Here, however, discarding the eigenvectors corresponding to the smallest eigenvalue helps in calculating the true similarity even when the patches are noisy.

One may easily verify this from the following example. Consider the vectors $v_p = [4, 5, 6, 7, 6, 5, 4]^T$, $v_r = [4, 4, 6, 8, 6, 4, 4]^T$ and $v_s = [1, 2, 3, 4, 3, 2, 1]^T$. In SVD domain representation of a matrix that has these vectors as its columns, the rows of matrix $V \Sigma$ are obtained to be $w_1 = [-14.21, 0.89]$, $w_2 = [-14.12, -0.41]$ and $w_3 = [-6.52, -1.05]$, while $cos(\theta_{pr}) = 0.9958$ and $cos(\theta_{ps}) = 0.9756$. On the other hand, if we directly use the vectors $v_p, v_r, v_s$ instead of $w_1, w_2, w_3$, respectively, we get $cos(\theta_{pr}) = 0.9926$ and $cos(\theta_{ps}) = 0.9734$. Thus, it may be noted that unlike calculating the correlation directly between the actual patches, our method performs the similarity comparison in the SVD domain wherein by discarding the smallest eigenvalue and the associated eigenvector, the obtained similarity values are robust to noisy patches.

The comparison of patches $\Phi_r$ and $\Phi_s$, which overlap with and are row, column adjacent to the patch $\Phi_p$, enables us to simultaneously capture horizontal, vertical and diagonal discontinuities. We now create a similarity matrix $S$ such that its element $S(p)$ represents the average similarity value of patch $\Phi_p$ with overlapping patches $\Phi_r$ and $\Phi_s$ calculated as follows,

$$S(p) = \frac{1}{2} \left[ \cos(\theta_{pr}) + cos(\theta_{ps}) \right], \quad \forall p \equiv (x, y) \in I_V. \tag{8.9}$$

## 8.2.2  THRESHOLDING

Once the similarity matrix $S$ is obtained, the idea is to detect the cracks by thresholding the values of $S$ which may have different values for different input images. It may be noted that if the overlapping patches in an input image have very high similarity, then the corresponding matrix $S$ may have many values that are nearer to 1, and therefore a high value of threshold $\delta$ could be required for correctly detecting the patches that have discontinuities that represent the cracks. Therefore, one has to choose the threshold $\delta$ based on the input image for correct detection of cracked regions, which we describe as follows.

In order to select the threshold value $\delta$ dynamically for a given image, we consider three quantities derived from the similarity matrix $S$, viz. the average value $avg(S)$, minimum value $min(S)$ and the maximum value $max(S)$. Since the compared patches are adjacent and also overlap each other, they have a high amount of similarity in their content. Therefore, it is reasonable to assume that the values in $S$ less than the average value $avg(S)$ would definitely correspond to the patches having discontinuities and hence represent cracks. Thus, the lowest value that $\delta$ may take is $avg(S)$.

If the difference between lowest and highest values of $S$ is high, it suggests that the values corresponding to patches with discontinuities are spread over a wider range, whereas the spread is over a narrow range when the difference is small. If the values in the similarity matrix $S$ vary in a narrow range, then the threshold value $\delta$ would be nearer to $avg(S)$. Thus, we infer that the threshold value has to be higher than the average value $avg(S)$ and also should depend on the minimum $min(S)$ and maximum $max(S)$ values. We set an initial threshold $\alpha$ to be an average of these three terms as given in the following Equation (8.10).

$$\alpha = \frac{min(S) + max(S) + avg(S)}{3}.$$

(8.10)

However, experimentally we found that a correction factor depending on the value $\alpha$ is required for correct detection of the cracks. Based on our experimentation, we arrive at the following Equation (8.11) that incorporates suitable correction factors to determine the threshold $\delta$.

$$\delta = \begin{cases} \alpha + 0.10, & \text{if } 0 \leq \alpha < 0.90, \\ \alpha + 0.05, & \text{if } 0.90 \leq \alpha < 0.95, \\ \alpha + 0.01, & \text{if } 0.95 \leq \alpha < 0.99, \\ \alpha, & \text{if } \alpha \geq 0.99. \end{cases}$$

(8.11)

In this way, the initial threshold $\alpha$ is calculated automatically, based on which an appropriate correction factor is added, to dynamically set the threshold $\delta$ depending on the input image.

Once we obtain the threshold $\delta$, we use it to detect the cracks by thresholding the values in the matrix $S$ representing the similarity of patches in the *intensity* image $I_V$. If $S(p) < \delta$, we declare the corresponding patches $\Phi_p$, $\Phi_r$ and $\Phi_s$ to be significantly dissimilar. Using this criteria, all the values in the matrix $S$ are compared with threshold $\delta$ to detect dissimilar patches, using which a binary image $B$ is constructed as,

$$B(\Phi_p) = \begin{cases} 1, & \text{if } S(p) \geq \delta, \forall p \equiv (x, y) \in I_V, \\ 0, & \text{otherwise,} \quad \text{and} \end{cases}$$

$$B(\Phi_s) = B(\Phi_r) = B(\Phi_p). \tag{8.12}$$

The binary image $B$ generated in this way by thresholding has the cracked regions represented by the value 1. Once the cracks are detected, an existing inpainting technique can be used to inpaint them. For illustration purpose we use the exemplar-based inpainted technique proposed in Criminisi et al. [21] to inpaint the detected cracks. In the following section, we illustrate the working of the SVD-based approach by using few experimental results.

### 8.2.3  EXPERIMENTAL RESULTS

We now discuss the results of our experiments on images downloaded from the Internet [44], as well as on those captured by us. In our experiments, the size of patches $\Phi_p$, $\Phi_r$ and $\Phi_s$ is set to $3 \times 3$. As mentioned earlier in Section 8.2.2, the detected cracks are inpainted using the technique proposed in Criminisi et al. [21] in order to demonstrate the suitability of our method to auto-detect cracked regions for inpainting.

The results of our experiments on wall and ceiling images are shown in Figure 8.10, and those on images pavements are illustrated in Figure 8.11. The input images of wall and ceiling are shown in Figure 8.10a. Since the ground truth containing the marked cracked regions was not available for the input images, we have considered regions marked by volunteers as cracks for comparison. The images containing the regions marked as cracks by volunteers are shown in Figure 8.10b. The detected cracks using our method are shown in Figure 8.10c and the corresponding inpainted images are shown in Figure 8.10d. Here the input image in the first row contains a crack having high contrast with respect to its surroundings. Although this appears to be a trivial case, one may note that the presence of tiny dark regions over the image makes the detection a challenging task. The detection performed by our method is similar to the region marked by the volunteers. The input image in the second row contains cracks along with other damaged regions. In this case also our method well detects the cracked region. A more complex case containing the crack in an image having low contrast is shown in the input image in the third row. Even in this case our method performs better. Likewise, in the pavement images shown in Figure 8.11, the detected cracked regions are similar to those marked by the volunteers, indicating the efficacy of our crack detection method.

In order to quantify the accuracy of the detected cracks, once again we consider the popularly used recall and precision metrics [140] defined in Equation (8.7). The performance in terms of *Recall* and *Precision* values for input images in Figures 8.10 and 8.11 is given in Table 8.3. The crack detection results obtained using our technique are significantly similar to the detection performed manually by volunteers which is also evident from Table 8.3. We observe that *Recall* value

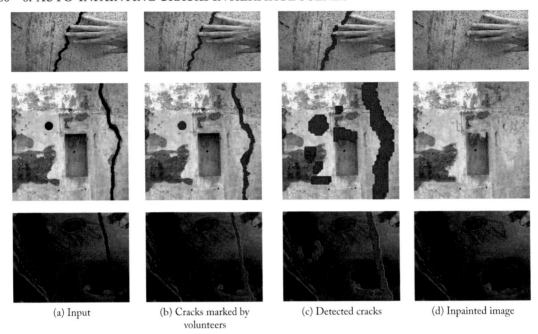

(a) Input  (b) Cracks marked by volunteers  (c) Detected cracks  (d) Inpainted image

**Figure 8.10:** Detection of cracks in images of wall and ceiling. (a) Input images; (b) cracks marked by volunteers are shown in red color; (c) cracks detected using our method; (d) inpainting of the detected cracks in (c) using the technique proposed in Criminisi et al. [21].

**Table 8.3:** Performance of the proposed technique in terms of Recall and Precision for the results shown in Figures 8.10 and 8.11

| Input | # Target Pixels | Recall | Precision |
|---|---|---|---|
| Image in row 1 of figure 8.10 | 5,414 | 0.9540 | 0.6702 |
| Image in row 2 of figure 8.10 | 2,513 | 0.9988 | 0.2290 |
| Image in row 3 of figure 8.10 | 5,431 | 0.9742 | 0.3708 |
| Image in row 1 of figure 8.11 | 40,741 | 0.9914 | 0.4445 |
| Image in row 2 of figure 8.11 | 5,613 | 0.9984 | 0.4772 |
| Image in row 3 of figure 8.11 | 29,333 | 0.8919 | 0.5781 |

for all the detected cracks in these images is nearer to 1. This clearly indicates that the desired pixels of the cracked regions have been detected.

We now show results on images captured by our camera in Figure 8.12, that include indoor and heritage scenes. These results are compared with the defect detection technique proposed

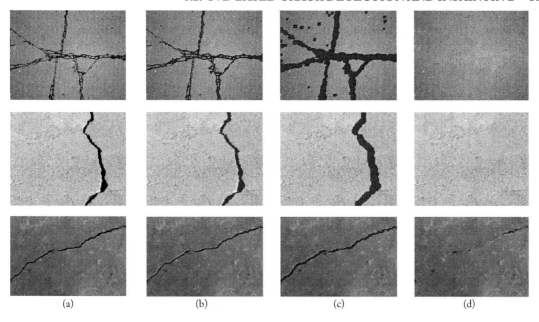

(a)                    (b)                    (c)                    (d)

**Figure 8.11:** Detection of cracks in pavement images. (a) Input images; (b) cracks marked by volunteers are shown in red color; (c) cracks detected using our method; (d) inpainting of the detected cracks in (c) using the technique proposed in Criminisi et al. [21].

**Table 8.4:** Performance comparison in terms of Recall and Precision for the images shown in Figure 8.12

| Input | # Target Pixels | Proposed Technique | | | Technique in [3] | | |
|---|---|---|---|---|---|---|---|
| | | Recall | Precision | Time (sec) | Recall | Precision | Time (sec) |
| Image in row 1 | 8,217 | 1.0000 | 0.5372 | 4.63 | 0.1503 | 0.0093 | 512 |
| Image in row 2 | 1,353 | 1.0000 | 0.1749 | 4.41 | 0.6438 | 0.0260 | 093 |
| Image in row 3 | 3,494 | 0.9531 | 0.2971 | 4.51 | 0.0000 | 0.0000 | 109 |

in Amano [3]. It may be noted that the results for technique Amano [3] are obtained after finetuning the parameters while the parameters $\alpha$ and $\delta$ in our method are dynamically calculated, depending on the input image. The performance comparison for these results in term of recall and precision is shown in Table 8.4. From the results shown in Figure 8.12 and the performance comparison in Table 8.4, it is clear that the desired cracked regions are successfully detected by our method. Although the technique in Amano [3] is good for detection for an alteration to the photograph like overlay text, our method is comparatively fast and more suitable when it comes to detection of damage in the photographed scene/object. In comparison to our previous crack

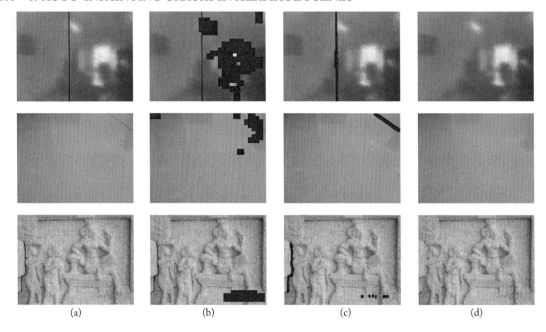

     (a)           (b)           (c)           (d)

**Figure 8.12:** Detection of cracks in indoor and heritage scene images captured by us. (a) Input images; (b) cracks detected using the technique proposed in Amano [3] are shown in red color; (c) cracks detected using our method are shown in red color; (d) inpainted image for the cracks detected by our method.

detection approach, the recall using the proposed approach is more or less similar but the precision shows improvement indicating higher accuracy in detecting the cracks. This can be observed from Tables 8.2 and 8.4 for the results on an image of a heritage site shown in row 1 of Figure 8.6 and row 3 of Figure 8.12.

## 8.3    CRACK DETECTION USING TOLERANT EDIT DISTANCE AND INPAINTING

Having discussed an SVD-based approach for crack detection in Section 8.2, we now discuss another approach based on tolerant edit distance measure to enhance the accuracy of detection of cracks. Cracks are typically characterized by dark areas in an image which can be easily identified by human beings but pose difficulty to computers. In trivial cases, simple thresholding is sufficient for detecting the cracks. However, as seen in the previous Sections 8.1–8.2, the subtle variations in pixel intensities make the detection of cracks a challenging task.

     The approach that we now discuss uses similarity of non-overlapping adjacent patches as a cue to localize the cracked regions. The novelty here is that we compare the patches for similarity

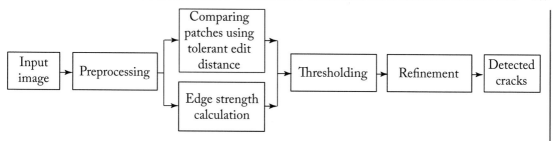

**Figure 8.13:** Our approach for crack detection using tolerant edit distance.

using a measure derived from the edit distance, which is a popular string metric used in the area of text mining. This helps in better localization of the cracks since the patch similarity is determined by avoiding a strict pixel-to-pixel comparison inside the patches that was done in our previous approach. Further, the refinement of the localized cracked regions is performed using a sophisticated segmentation technique unlike our previous approach which did not perform refinement. The overall process in the proposed approach thus provides the cracked regions detected in a more accurate manner. An existing inpainting technique is then used to perform the digital repair of the detected cracked regions. A block diagram of this approach for crack detection is shown in Figure 8.13 the details of which are discussed below.

### 8.3.1 PREPROCESSING

For a given input image $I$ of size $M \times N$ we perform a pre-processing step by considering its intensity normalized version $I_0$. Since the cracked regions are dark, the low-intensity pixels are more likely to be part of a crack. We construct a weight matrix $I_w$ from $I_0$ such that dark pixels have higher weights given by,

$$I_w(x, y) = \exp(-I_0(x, y)), \qquad (8.13)$$

where $(x, y)$ denote the pixel coordinates. The weights in $I_w$ are multiplied with the corresponding pixels in $I_0$ and the resulting image is eroded to obtain an image $I_v$ that we use for further processing. The erosion operation is performed so that the narrow dark regions are enlarged for proper detection, which may otherwise remain undetected in later operations. The preprocessing step is illustrated with an example in Figure 8.14.

### 8.3.2 PATCH COMPARISON USING TOLERANT EDIT DISTANCE

After the preprocessing step, we obtain the similarity of patches of size $m \times n$ in $I_v$ with their respective right and bottom neighbors. Note that unlike the previous method, here we use non-overlapping patches. Thus, for a patch $\Phi_p$ at pixel $p$ with coordinates $(x, y)$ in the image $I_v$, the right and bottom non-overlapping patches considered for similarity are $\Phi_r$ and $\Phi_s$ at pixels

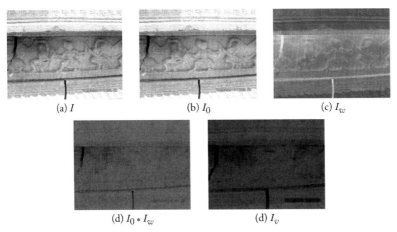

(a) $I$          (b) $I_0$          (c) $I_w$

(d) $I_0 * I_w$          (d) $I_v$

**Figure 8.14:** Preprocessing of an input image.

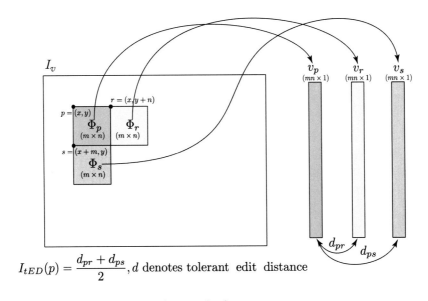

$$I_{tED}(p) = \frac{d_{pr} + d_{ps}}{2}, d \text{ denotes tolerant edit distance}$$

**Figure 8.15:** Patch comparison using tolerant edit distance.

$r = (x, y + n)$ and $s = (x + m, y)$ as shown in the Figure 8.15. Let the pixels of patches $\Phi_p$, $\Phi_r$ and $\Phi_s$ be rearranged using lexicographical ordering to form vectors $v_p$, $v_r$ and $v_s$, respectively.

A simple method for comparison of the non-overlapping patches is to calculate the sum of absolute difference or sum of squared difference (SSD) across corresponding pixels of the compared patches. These measures are, however, sensitive to noise and may give a high error even for visually similar patches, which is evident in Figure 8.16. Moreover, comparing a patch with its

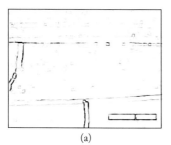 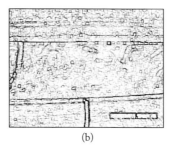 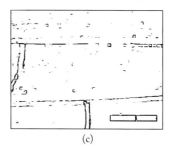

(a)                                (b)                                (c)

**Figure 8.16:** Comparison of (a) sum of absolute difference image, (b) sum of squared difference image, and (c) tolerant edit distance image $I_{tED}$ for tolerance $\delta_t = 10$. Patch size is $3 \times 3$. With the input image of size $684 \times 912$ we have $I_{tED}$ of size $227 \times 303$. Here an enlarged, intensity-inverted version is shown for clarity.

spatially shifted version also gives high error, where in fact both are visually identical. Thus, it becomes difficult to separate the cracks from its neighborhood using a threshold.

In order to overcome this drawback, we instead use a measure derived from the edit distance [126] along with a tolerance value. The edit distance is robust to spatial shifts and the use of the tolerance value overcomes the effect of noise in the compared patches. Thus, we measure the similarity between patches $\Phi_p$, $\Phi_r$ and $\Phi_s$ by calculating the tolerant edit distance (tED) $d_{pr}$ and $d_{ps}$ between the pairs $v_p, v_r$ and $v_p, v_s$, the average of which is assigned to the pixel $p$. An algorithm for computing the tED is given in Algorithm 8.1. The tED calculated using Algorithm 8.1 is such that pixel values within a tolerance $\delta_t$ are considered to be equivalent. It is calculated for all the patches for which there exist both left and bottom non-overlapping adjacent patches. The calculated tED values are used to form an image $I_{tED}$, which, when multiplied with an edge strength image makes it easier to detect the cracked regions.

## 8.3.3   EDGE STRENGTH CALCULATION

Since the cracked regions are distinct from their neighboring regions, they exhibit higher edge strengths. In order to give preference to patches having higher edge strengths, we generate an image $I_g$ consisting of normalized gradient magnitudes from the preprocessed image $I_v$. Now, by convolving the image $I_g$ with horizontal, vertical, diagonal and anti-diagonal line filters of size $3 \times 3$ with filter masks shown in Figure 8.17, the maximum response at each pixel is recorded to create an image $I_m$. This image is updated by discarding the low responses followed by morphological closing to detect the connected components.

The gradient magnitude image $I_g$ is now updated using the updated image $I_m$, such that the highest gradient magnitude within each connected component is assigned to all the pixels within the respective component. Updating $I_g$ in this manner enables us to assign a unique edge strength value to distinct components. The edge strength image $I_e$ is now constructed by taking

**Algorithm 8.1** Calculation of tED

% For vectors $v_1$ and $v_2$ with lengths $|v_1|$ and $|v_2|$, respectively and $\delta_t$ as tolerance value,

% Initialization
$D[0,0] := 0$
**for** $i := 1$ to $|v_1|$ **do** $D[i,0] := i$ **end for**
**for** $j := 1$ to $|v_2|$ **do** $D[0,j] := j$ **end for**

% Required operation: substition, insertion or deletion
**for** $i := 1$ to $|v_1|$ **do**
   **for** $j := 1$ to $|v_2|$ **do**
      $m_1 := D[i-1, j-1] + C(v_1[i], v_2[j], \delta_t)$
      $m_2 := D[i-1, j] + 1$
      $m_3 := D[i, j-1] + 1$
      $D[i,j] = \min(m_1, m_2, m_3)$
   **end for**
**end for**

% Result
**return** $\text{tED} := D[|v_1|, |v_2|]$

% Comparison function: $C(v_1[i], v_2[j], \delta_t)$
**if** $||v_1[i] - v_2[j]|| \leq \delta_t$ **then**
   $C(v_1[i], v_2[j], \delta_t) := 0$
**else**
   $C(v_1[i], v_2[j], \delta_t) := 1$
**end if**

the normalized sum of $I_g$ and $I_w$. We now multiply these edge strengths with the corresponding tED, to get the weighted tED image $I_{tw}$. This process is illustrated with an example in Figure 8.18.

In order to fill the gap between the boundaries, a morphological closing operation is applied on $I_{tw}$, with the size of the structuring element depending on the size of the preprocessed image $I_v$. The morphologically closed image $I_{tw}$ is now multiplied with the resized version of the weight matrix $I_w$ to obtain an intermediate image $I_{wc}$. In order to assign unique values to different objects for segmentation in image $I_{wc}$, we employ the method used earlier for updating the gradient magnitude image $I_g$ described in the previous paragraph. Thus, by convolving the intermediate image $I_{wc}$ with the 3 × 3 line filters, thresholding the maximum response image and applying the morphological closing operation, we obtain the image $I_c$ of size $(\frac{M}{m} - 1) \times (\frac{N}{n} - 1)$, as shown in Figure 8.19a in which the connected components have unique values.

| 0 | 0 | 0 |
|---|---|---|
| 1 | 1 | 1 |
| 0 | 0 | 0 |

| 1 | 0 | 0 |
|---|---|---|
| 0 | 1 | 0 |
| 0 | 0 | 1 |

| 0 | 1 | 0 |
|---|---|---|
| 0 | 1 | 0 |
| 0 | 1 | 0 |

| 0 | 0 | 1 |
|---|---|---|
| 0 | 1 | 0 |
| 1 | 0 | 0 |

(a)                (b)                (c)                (d)

**Figure 8.17:** Line filters. (a) Horizontal, (b) main diagonal, (c) vertical, and (d) anti-diagonal.

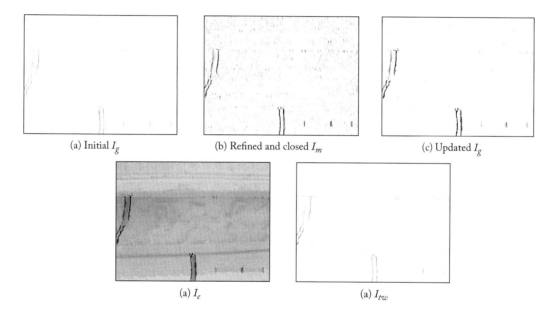

(a) Initial $I_g$          (b) Refined and closed $I_m$          (c) Updated $I_g$

(a) $I_e$                    (a) $I_{tw}$

**Figure 8.18:** Edge strength $I_e$ and weighted tolerant edit distance images $I_{tw}$. Sizes of $I_g$, $I_m$ and $I_e$ are the same as that of $I_0$, while $I_{tw}$ and $I_{tED}$ are of the same size. Here an enlarged and intensity-inverted version of $I_{tw}$ is shown for clarity.

## 8.3.4 THRESHOLDING

The higher the value of a region in the image $I_c$, the more likely it is to be a crack. Thus, the regions with values lower than a threshold $T$ need to be discarded. Let $V$ denote the array consisting of $k$ unique values in $I_c$ arranged in ascending order. Then, inspired by the threshold selection method for matching features of the scale invariant feature transform (SIFT) given in Lowe [72], we estimate the threshold $T$ using Algorithm 8.2.

The image $I_c$ is now updated by setting values less than $T$ to zero. Each pixel in $I_c$ corresponds to an $m \times n$ overlapping patch in $I_v$. We obtain an initial crack-detected image $I_1$, which is of the same size as that of $I_v$, by copying pixels values from $I_c$ to corresponding patches in $I_1$.

**Algorithm 8.2** Selection of threshold $T$

% For an array $V$ consisting of $k$ unique values in $I_c$ arranged in ascending order,

% Initialize
$T := V[k]$

% Update
**for** $i := k - 1$ to 1 **do**
  **if** $V[i] < 0.2$ **then**
    **break**
  **end if**
  **if** $(\frac{V[i]}{V[i+1]}) \geq (\frac{V[i-1]}{V[i]})$ **then**
    $T := V[i]$
  **end if**
**end for**

% Result
**return** $T$

A second morphological closing operation is now applied on the binary image $I_1$ in order to avoid splitting of the detected region. Note that the image $I_1$ as shown in Figure 8.19b gives a good estimate of the cracked regions, however, a few pixels of the cracked regions which are similar to the surroundings may still remain undetected. Therefore, a refinement step is required to achieve a more accurate detection.

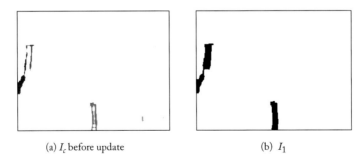

(a) $I_c$ before update          (b) $I_1$

**Figure 8.19:** Initial detection. Image $I_c$ is thresholded and mapped to $I_v$ to obtain $I_1$. Size of $I_c$ is the same as that of $I_{tED}$, while $I_1$ and $I_v$ are of the same size. Here an enlarged, intensity-inverted version of $I_c$ is shown for clarity.

### 8.3.5    REFINEMENT

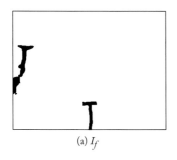

(a) $I_f$                                (b) $I_f$ overlapped on input image                        (c) Inpainted image

**Figure 8.20:** Refinement of the initially detected cracks. (a) Final detection binary image $I_f$, (b) detected regions overlapped on the input image, and (c) inpainted result.

Interactive image segmentation techniques based on curve evolution [14] and graph-cut optimization [103] have been widely used for accurately detecting roughly marked objects. For refining $I_1$, we use the method based on active contours[1] proposed in Chan and Vese [14], to obtain the final crack-detected binary image $I_f$, an example of which is shown in Figure 8.20a. Here we use automatically detected $I_1$ in place of the interactive input for segmentation. In order to justify the suitability of the proposed method for inpainting, we also show the inpainted result in Figure 8.20c obtained using the method proposed in Criminisi et al. [21].

### 8.3.6    EXPERIMENTAL RESULTS

In our experiments, we show the results for three input images of size $684 \times 912$ captured from the world heritage site in Hampi, Karnataka, in India. We considered patches $\Phi_p$ of size $3 \times 3$ in our experiments. Use of patches having larger sizes did not significantly improve the detection. In calculation of the tolerant edit distance, we have set the tolerance value $\delta_t = 10$ based on the following experimentation. We considered many patches at the boundary of known cracked regions from a number of images, along with their corresponding non-overlapping adjacent patches. For each of these patches, we calculated the tolerant edit distances by varying values of $\delta_t$. Curves of tolerant edit distance vs. probability of patches, corresponding to every $\delta_t$ were plotted, as shown in Figure 8.21. Since the patches belonged to crack boundaries, we have higher edit distance (i.e., $\delta_t = 0$). Increasing the value of $\delta_t$ reduces the sensitivity, and therefore only large variations can be detected. It is observed that for $\delta_t = 10$, sufficiently large variations were detected and further increasing $\delta_t$ did not change the curve significantly. The size of structuring element for morphological closing used for filling in large gaps depends on the image size. For an image of size

---

[1]For active contour segmentation technique, we have used the implementation available at http://www.mathworks.in/mat
labcentral/fileexchange/23847-sparse-field-methods-for-active-contours

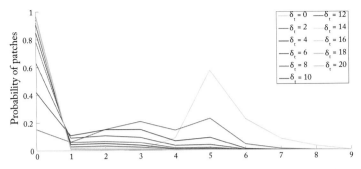

Tolerant Edit Distance (tED) for a set of 3 × 3 patches on crack boundary

**Figure 8.21:** Curves for varying tolerance values $\delta_t$.

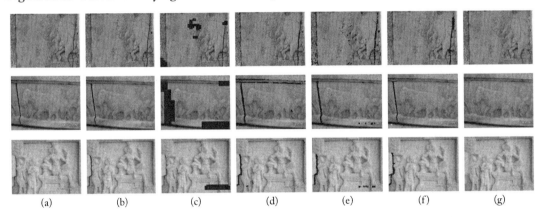

**Figure 8.22:** Comparison of crack detection techniques: (a) input image; (b) manual selection by volunteers; detection results; (c) Amano [3], (d) Turakhia et al. [121], (e) Padalkar et al. [90], (f) proposed method, (g) inpainted image using regions detected in (f).

$M \times N$, the size of structuring element is chosen as $(\max(M, N)/360 + \min(M, N)/270)$ such that $M, N > 270$.

In Figure 8.22 we show a comparison of our results with those obtained using the techniques in Amano [3], Turakhia et al. [121] and our SVD-based crack-detection method Padalkar et al. [90] discussed in Section 8.2. It may be noted that the results for the technique in Amano [3] are obtained after fine-tuning the parameters. We show the regions marked as cracks by volunteers in Figure 8.22b. Inpainted results by considering the cracked regions detected by the proposed method are shown in Figure 8.22g to demonstrate its suitability for automating the repair of cracks.

To determine the suitability of the resulting detection for use by inpainting algorithms, the popularly known recall and precision metrics defined in Equation (8.7) are considered. However, for showing an insight of the robustness of our proposed algorithm, we use a slightly different precision measure defined as $Precision = \frac{|Ref_{conn}|}{|Dect|}$. Here $Dect$ are the pixels detected by the algorithm to be in the cracked regions and $Ref_{conn}$ are those pixels detected in $Dect$ that are connected to cracked regions marked by volunteers.

The quantitative measures recall and precision for images displayed in Figure 8.22 are given in Table 8.5. From the table we observe that our method performs better crack detection. Also, in comparison to our previous crack-detection approaches, the recall values for the proposed approach are similar, but the precision values show significant improvement indicating higher accuracy of detection. Thus we have progressed from a primitive pixel-based approach for crack-detection in Section 8.1 to a patch-based method in Section 8.2 and further to a sophisticated patch-based approach that we discussed above, each showing improvement in the accuracy of detecting the cracks. In the next section we extend this approach to perform automatic inpainting in the videos captured at heritage sites.

Table 8.5: Comparison in terms of recall and precision for images in Figure 8.22

| Input | Amano [3] | | Turakhia et al. [121] | | Padalkar et al. [90] | | Proposed Approach | |
|---|---|---|---|---|---|---|---|---|
| | Recall | Precision | Recall | Precision | Recall | Precision | Recall | Precision |
| Image in row 1 | 0.046 | 0.068 | 0.749 | 0.678 | 0.863 | 0.392 | 0.840 | 0.997 |
| Image in row 2 | 1.000 | 0.579 | 0.974 | 0.974 | 0.987 | 0.857 | 0.985 | 0.996 |
| Image in row 3 | 0.000 | 0.000 | 0.988 | 0.887 | 0.953 | 0.743 | 0.990 | 1.000 |

## 8.4   EXTENSION TO AUTO-INPAINT CRACKS IN VIDEOS

For extending the proposed approach of crack detection and inpainting to videos, one may think of performing frame-by-frame detection and inpainting. This abstraction, however, in practice is a long-drawn-out process, as it does not exploit the inter-frame redundancy. Also, there may be occlusion or change in illumination across frames as the camera moves, due to which the pixels corresponding to cracked regions detected in one frame may not map to the pixels corresponding to the same cracked regions detected in some other frame. Since the inpainting task is highly sensitive to the pixels to be inpainted, it leads to large variations in the two inpainted frames for the same cracked regions. As a result, the auto-inpainted videos created by detecting and inpainting frame-by-frame, appear unstable and the effect of seam becomes visible.

Alternatively, one may use motion as a cue to track and inpaint the cracked regions across subsequent frames. Motion estimation and compensation have been popularly used in video compression techniques [56, 112]. Here intermediate frames are generated using independent frames and motion parameters. However, since these methods are block based, their use to inpaint videos results in blocking artefacts. Moreover, they are computationally expensive and the motion parameters are estimated using 2D-2D transformation. Frame-to-frame transformations are, therefore, needed to track the damaged regions in subsequent frames for creating a seamlessly inpainted video.

Brown and Lowe [9] have suggested a method for automatic image stitching, wherein transformation between the images to be stitched is calculated by matching keypoints invariant to rotation, scaling and view point. Here the transformation is considered to be projective or a homography [49]. Since the videos captured at heritage sites usually contain nearly planar rigid objects/scene with a moving camera, we can consider the video frames to be images captured from different viewpoints. Hence the transformation between these frames can be represented by a homography.

In our video inpainting method, we consider pairs of temporally adjacent frames and use the homography [49] to track the cracked regions from one frame to another. The first video frame is initially considered as the reference frame, which is later updated based on the camera movement. The cracked regions are detected in reference frames using the method described in Section 8.3 and then tracked to subsequent frames. Similarly, the detected cracks are inpainted in the reference frames using the technique proposed in Criminisi et al. [21] and then mapped to the tracked regions in the subsequent frames. Note that the inpainting of video frames cannot be done by simply copying objects visible in other frames, as done in Patwardhan et al. [95]. This is because an object to be inpainted in one frame also needs to be inpainted in other frames as well, which mandates the use of a hole-filling technique. Our approach for detecting and inpainting the cracked regions in videos is shown in Figure 8.23. In what follows, we briefly describe the important stages involved in automating the inpainting in videos viz. (1) estimation of homography, (2) reference frame detection and (3) tracking and inpainting the cracked regions across frames.

### 8.4.1   HOMOGRAPHY ESTIMATION

As already mentioned, since the videos captured at heritage sites usually contain nearly planar rigid objects/scene with a moving camera, we consider the video frames to be images captured from different viewpoints. Hence the transformation between these frames can be represented by a homography [49, 65], which we estimate by extracting keypoints and matching the scale invariant feature transform (SIFT) descriptors[2] [72].

The estimation of homography is done by a random sampling consensus (RANSAC) [34] of all the matching keypoints at locations $(x_1, y_1)$ and $(x_2, y_2)$ in the two frames that obey the relationship $[x_2, y_2, 1]^T = H[x_1, y_1, 1]^T$, where $H$ is a $3 \times 3$ non-singular matrix representing

---

[2]An implementation for extraction and matching of SIFT keypoints and corresponding descriptor is available at http://www.cs.ubc.ca/~lowe/keypoints/

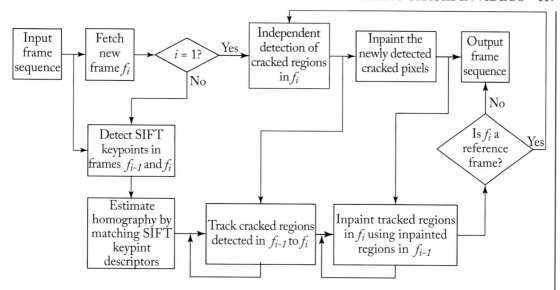

**Figure 8.23:** Our approach for extending crack detection and inpainting in images to videos of heritage scenes.

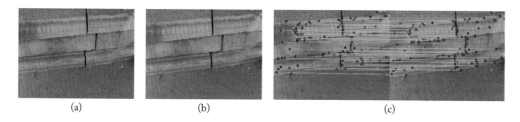

**Figure 8.24:** Matching of SIFT keypoints. (a)–(b) Two frames of a video; (c) pairs of matching keypoints shown by green joining lines.

the homography.[3] Figure 8.24 illustrates the matching of SIFT keypoints between a pair of video frames.

## 8.4.2 REFERENCE FRAME DETECTION

While capturing the video with a moving camera, new cracked regions may appear. Therefore, an independent crack detection needs to be performed quasi-periodically depending on the camera movement. An intuitive way to quantify the camera motion is to calculate the magnitude of translation. Faugeras and Lustman [33], Ma et al. [77] have shown that, given a homography

---

[3]For fitting homography to keypoints using RANSAC, we used the code available at http://www.csse.uwa.edu.au/~pk/Research/MatlabFns/Robust/ransacfithomography.m

matrix, it can be decomposed to estimate the translation. The decomposition yields four solutions in general out of which only two are physically possible. However, each of these solutions has the same magnitude of translation. We make use of this information to detect the reference frame. The solutions for decomposition[4] of a homography $H$ are obtained using the method in Ma et al. [77].

Let $t$ be the translation vector of one of the obtained solutions such that $t = [t_1, t_2, t_3]^T$. Then the magnitude of translation is given by $|t| = \sqrt{t_1^2 + t_2^2 + t_3^2}$. Also, let $\delta_r$ be the threshold for translation. Considering the first video frame as a reference $ref$, a homography along with the translation between the reference and every incoming frame $f_i$ is calculated. If the magnitude of translation is above a threshold $\delta_r$, then the incoming frame $f_i$ is declared to be a reference frame.

For selection of the translation threshold $\delta_r$ to detect the reference frames, we conducted the following experiment. We manually selected two frames viz. (1) the frame in which a cracked region has completely appeared and (2) the frame in which the next cracked region begins. For every such pair of frames, translation was calculated. Conducting the experiment on a number of videos having frames of size $270 \times 360$ revealed that the average value of $\delta_r = 25$ can be used to detect new incoming cracked regions. However, the problem with this threshold is that, while a part of the newly appearing cracked region gets detected successfully, the remaining part which appears in subsequent frames is not detected. For successful detection of the complete cracked regions, a lower value of threshold is required. By keeping the threshold from 25 to 0, we found $\delta_r = 5$ to be an appropriate threshold for successful detection of the complete cracked regions. Also note that the intensity change in corresponding pixels across the frames within this small translation is negligible. This enables a seamless copying of pixel values when propagating the already inpainted cracked regions across subsequent frames. We have, therefore, set $\delta_r = 5$.

### 8.4.3   TRACKING AND INPAINTING CRACKED REGIONS ACROSS FRAMES

For a pair of temporally adjacent frames $f_{i-1}$ and $f_i$, the locations $(x_i, y_i)$ of cracked pixels in $f_i$ can be tracked from the frame $f_{i-1}$ using the corresponding locations $(x_{i-1}, y_{i-1})$ as follows,

$$
\begin{bmatrix} x_i' \\ y_i' \\ z_i' \end{bmatrix} = H_i \begin{bmatrix} x_{i-1} \\ y_{i-1} \\ 1 \end{bmatrix}, \tag{8.14}
$$

where $(x_i', y_i', z_i')$ are the homogeneous coordinates for the point $(x_i, y_i)$ such that $x_i = \frac{x_i'}{z_i'}$, $y_i = \frac{y_i'}{z_i'}$ and $H_i$ denotes the homography between frames $f_{i-1}$ and $f_i$. Here it may happen that

---

[4]For decomposition of estimated homography, we have used the implementation available at http://cs.gmu.edu/~koseck a/examples-code/homography2Motion.m

estimated coordinates $x_i$ and $y_i$ are real numbers. These are rounded to the nearest integers so that we have the tracked pixels at integer locations. For simplicity, let the integer-rounded location coordinates be denoted by $(x_i, y_i)$. Setting these cracked pixel locations to 1 with all other locations set to a value of 0, a crack-mask consisting of 1's and 0's is constructed for the frame $f_i$. Since homography introduces geometric distortions, it may happen that some narrow cracked regions detected in the frame $f_{i-1}$ may become disjoint regions in the newly constructed crack-mask, which leads to some part of the cracked regions being missed out. In order to avoid this, we use morphological closing on the crack-mask to connect the nearby disjoint regions. The crack-mask now gives the locations of the tracked cracks in the frame $f_i$ that correspond to the crack regions detected in the frame $f_{i-1}$.

We now describe how an incoming frame is processed. The first video frame $f_1$ being a reference frame is independently inpainted after identifying the cracked regions in it. Any subsequent incoming frame $f_i$ may or may not be a reference frame depending on the camera motion. For both cases, we use the above procedure along with Equation (8.14) to track cracked regions from $f_{i-1}$ to $f_i$. Let $P_i$ denote the binary image consisting of the cracked regions tracked from frame $f_{i-1}$ to frame $f_i$. In case $f_i$ is not a reference frame, it can be inpainted by filling up the location of the tracked crack pixels (i.e., $\{(x_i, y_i) | P_i(x_i, y_i) = 1\}$). This is achieved by simply copying the values of the corresponding pixels from the inpainted version of the previous frame $f_{i-1}$. Note that the frames are temporally adjacent and the change in intensity of corresponding pixels is negligible. Also note that the selected translation threshold $\delta_r$ is small enough so that the change in intensity of corresponding pixels across frames within this translation is also negligible. Thus, the copying of pixel values across subsequent frames does not introduce any seam.

Since the homography matrix $H_i$ is non-singular, its inverse $H_i^{-1}$ exists. Therefore, the crack pixels at locations $(x_i, y_i)$ and the corresponding locations $(x_{i-1}, y_{i-1})$ from the previous frame $f_{i-1}$, must be related as follows,

$$\begin{bmatrix} x'_{i-1} \\ y'_{i-1} \\ z'_{i-1} \end{bmatrix} = H_i^{-1} \begin{bmatrix} x_i \\ y_i \\ 1 \end{bmatrix}, \tag{8.15}$$

where $(x'_{i-1}, y'_{i-1}, z'_{i-1})$ are the homogeneous coordinates for the point $(x_{i-1}, y_{i-1})$, such that $x_{i-1} = \frac{x'_{i-1}}{z'_{i-1}}$ and $y_{i-1} = \frac{y'_{i-1}}{z'_{i-1}}$. Since $x_i$ and $y_i$ were rounded to integers, we may obtain the corresponding $x_{i-1}$ and $y_{i-1}$ as real numbers. The intensity at this location is obtained by considering the first-order integer location neighborhood and using the bilinear interpolation. It may be noted that inpainting performed in this manner across is almost insensitive to small changes in the morphologically closed crack-mask due to directly copying the values from the previously inpainted regions.

If the incoming frame $f_i$ is a reference frame, then crack detection is performed independently. However, since only the newly appearing cracked pixels need to be inpainted, we first calculate the binary image $P_i$ consisting of the cracked regions tracked from the previous frame

$f_{i-1}$. Now, let $B_i$ denote the crack-detected binary image corresponding to $f_i$ obtained using the method described in Section 8.3. Then the binary image $Q_i$ consisting only of the newly appearing cracked pixels is given by,

$$Q_i(x_i, y_i) = \begin{cases} 1, & B_i(x_i, y_i) - P_i(x_i, y_i) > 0, \\ 0, & \text{otherwise.} \end{cases} \qquad (8.16)$$

Now an initial inpainting of the reference frame $f_i$ is achieved by using the inpainted version of the previous frame $f_{i-1}$ and the binary image $P_i$. The locations $(x_{i-1}, y_{i-1})$ in frame $f_{i-1}$ corresponding to the pixels at locations $\{(x_i, y_i) | P_i(x_i, y_i) = 1\}$ are obtained using the relation in Equation (8.15). Similar to inpainting a non-reference frame as described above, the pixels at locations $(x_i, y_i)$ are filled up by copying values from the corresponding pixels at locations $(x_{i-1}, y_{i-1})$ to obtain the initial inpainted image. The newly detected cracked pixels given by the binary image $Q_i$ are the holes to be filled up in the initially inpainted image. The final inpainted version of the reference frame is obtained by using the method proposed in Criminisi et al. [21] considering the initial inpainted image and the binary image $Q_i$ as inputs. An example for performing inpainting when a new reference frame appears is illustrated in Figure 8.25.

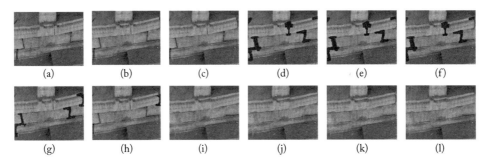

**Figure 8.25:** Inpainting a newly appearing reference frame $f_i$. (a), (b) and (c) show frames $f_{i-2}$, $f_{i-1}$ and $f_i$, respectively; the cracked regions corresponding to (a), (b) and (c) tracked from detected cracks in previous frames are shown in (d), (e) and (f); independent crack detection in $f_i$ is shown in (g), while the newly appearing cracked pixels in (g) with respect to (f) are displayed in (h); the inpainted versions of $f_{i-2}$, $f_{i-1}$, $f_i$ obtained by copying pixels from respective previous inpainted frames are shown in (i), (j) and (k); final inpainted version of $f_i$ obtained after inpainting the newly detected pixels is shown in (l). Note that the crack visible near the right side in (k) is filled in (l) by independently inpainting pixels shown in (h).

It may happen that a detected reference frame is highly blurred or noisy due to an unstable camera motion. In such a case, the crack-detection method described in Section 8.3 may fail and detect many regions as cracks. This can be avoided by simply thresholding the number of pixels in the newly detected cracked regions. Assuming that the number of pixels in the cracked regions do not vary substantially across the reference frames or whenever a new reference frame

---

**Algorithm 8.3** Video frame inpainting

---

% Let the $i^{th}$ video frame be denoted by $f_i$, such that the video consists of total $k$ frames. $R_i := 1$ denotes $f_i$ is a reference frame. Let $A_i$ denote the inpainted version of frame $f_i$.

% Initialization
Detect damaged regions in $f_1$ to get $B_1$.
Set threshold $\delta_0 := |B_1|$.
Perform inpainting on $f_1$ using $B_1$ to get $A_1$.

% Loop
**for** $i := 2$ to $k$ **do**
  Extract SIFT descriptors and homography $H_i$.
  Calculate $P_i$ by tracking damaged regions.
  **if** $R_i := 1$ **then**
    Detect damaged regions in $f_i$ to get $B_i$.
    Calculate $Q_i$ using $P_i$ and $B_i$.
    **if** $|Q_i| :\leq \delta_0$ **then**
      Calculate $S_i$ using $P_i$ and $B_i$.
      Fill pixels $\{(x_i, y_i)|S_i(x_i, y_i) = 1\}$ using $A_{i-1}$ to get initial inpainted image *init*.
      Perform inpainting on *init* using $Q_i$ to get $A_i$.
    **else**
      $R_i := 0, R_{i+1} := 1$.
      Fill tracked pixels $\{(x_i, y_i)|P_i(x_i, y_i) = 1\}$ using $A_{i-1}$ to get $A_i$.
    **end if**
  **else**
    Fill tracked pixels $\{(x_i, y_i)|P_i(x_i, y_i) = 1\}$ using $A_{i-1}$ to get $A_i$.
  **end if**
**end for**

% Result
**return** Inpainted frames $A_i \; \forall i := 1, \ldots, k$.

---

is encountered, we set a threshold $\delta_0$ based on the number of cracked pixels detected in the first frame. This is a valid assumption because, while the camera moves and new cracked regions enter a frame, some pixels of the previously detected cracked regions may exit. Also, even if the cracked pixels do not exit, we expect only a few new cracked pixels to enter. Therefore, the threshold $\delta_0$ accounting for the newly entering cracked pixels is set to half the number of cracked pixels detected in the first frame. Let $|Q_i|$ denote the number of newly detected cracked pixels in the frame $f_i$ and $|B_1|$ denote the number of cracked pixels detected in the first frame. Then, for a reference frame $f_i$, if we have $|Q_i| > \delta_0$ (such that $\delta_0 = 0.5 * |B_1|$), the frame $f_i$ is treated as a non-reference frame and inpainting is performed accordingly. Also, the frame $f_{i+1}$ is set as a reference frame, provided $f_i$ is not the last frame. The complete procedure for inpainting video frame is given in Algorithm 8.3.

### 8.4.4    EXPERIMENTAL RESULTS

We now present three results of our auto-inpainting method on videos captured by us from the heritage site in Hampi, Karnataka, India. Here we consider the camera movement to be unconstrained while capturing video of the heritage site, as such videos are typically captured by novices, hobbyists and tourists. The results are shown in Figures 8.26–8.28 where we show six frames of each video. Although the videos were captured at only one heritage site, our method is generic and should work for other heritage site videos. As an example, we show one more result on a video of the McConkie Ranch Petroglyphs near Vernal, Utah, in Figure 8.29 and it demonstrates the effectiveness of the proposed method. This video was uploaded by an enthusiast on the popular streaming site YouTube [108].

From the reported results, we can observe that by using our method, the detected cracked regions are effectively tracked and plausibly inpainted to get a seamless video. Although there exist approaches for semi-automatic inpainting of unwanted objects in videos [131] and video inpaint-

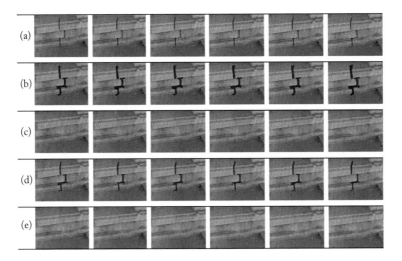

**Figure 8.26:** Result of auto-inpainting cracked regions in video frames. (a) Input frame sequence, left most frame is the reference frame; (b) cracked regions auto-detected in the reference frame tracked using SIFT and homography; (c) inpainted frames corresponding to frames in (b); (d) cracked regions auto-detected independently in each frame; (e) inpainted frames corresponding to frames in (d).

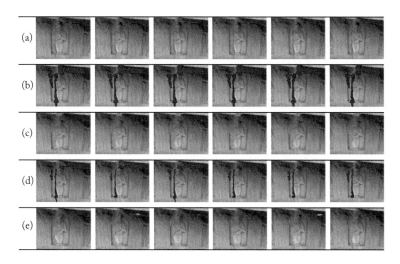

**Figure 8.27:** Result of auto-inpainting cracked regions in video frames. (a) Input frame sequence, left most frame is the reference frame; (b) cracked regions auto-detected in the reference frame tracked using SIFT and homography; (c) inpainted frames corresponding to frames in (b); (d) cracked regions auto-detected independently in each frame; (e) inpainted frames corresponding to frames in (d).

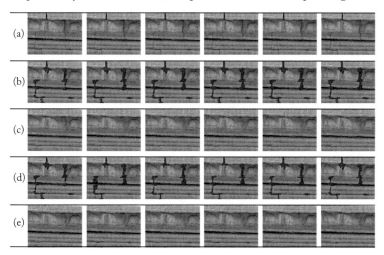

**Figure 8.28:** Result of auto-inpainting cracked regions in video frames. (a) Input frame sequence, left most frame is the reference frame; (b) cracked regions auto-detected in the reference frame tracked using SIFT and homography; (c) inpainted frames corresponding to frames in (b); (d) cracked regions auto-detected independently in each frame; (e) inpainted frames corresponding to frames in (d).

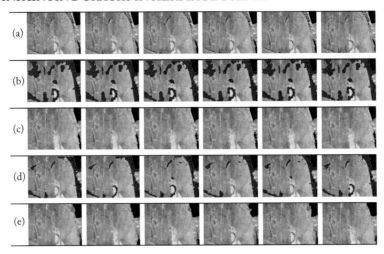

**Figure 8.29:** Result of auto-inpainting cracked regions in video frames. (a) Input frame sequence, left most frame is the reference frame; (b) cracked regions auto-detected in the reference frame tracked using SIFT and homography; (c) inpainted frames corresponding to frames in (b); (d) cracked regions auto-detected independently in each frame; (e) inpainted frames corresponding to frames in (d).

ing under constrained camera motion [95], it may be noted that, to the best of our knowledge, there does not exist any approach that demonstrates automatic video inpainting under unconstrained camera motion with no moving objects. We, therefore, do not show any comparison with the approaches in Patwardhan et al. [95], Wexler et al. [131]. However, we do compare the proposed approach with auto-inpainting done in a frame-by-frame fashion. The results of our method, along with auto-inpainting performed individually on every frame are shown in Figures 8.26–8.29.

For an objective comparison of our method with the frame-by-frame inpainting technique, we consider the no-reference video quality assessment (NR VQA) measures viz. blockiness and bluriness [30] and sudden local changes [105, 106]. These measures estimate the video quality directly from the processed (i.e., inpainted) video without considering the input video. Nevertheless, in an application like video inpainting, some information from the unprocessed video also can be used to quantify the quality of the processed video. Intuitively, to obtain a temporally plausible inpainted video, the optical flow of the input video should be maintained on inpainting, provided the objects to be inpainted are stationary. In other words, the optical flow between every pair of temporally adjacent frames in input and corresponding pair of frames in the inpainted video should be similar. The inpainting of only the stationary object is a valid assumption for inpainting videos of heritage monuments. With this cue, the optical flow between every pair of adjacent frames in both input as well as inpainted video can be estimated and used to quantify the quality of the inpainted video. One can estimate the optical flow by using the classic method proposed by Lucas and Kanade [74].

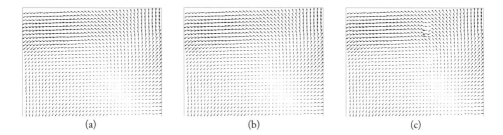

Figure 8.30: Optical flow between a pair of temporally adjacent frames in (a) input video, (b) auto-inpainted video using proposed method, (c) video generated by auto-inpainting every frame independently. The optical flow in (a) and (b) appear to be similar while some haphazard orientations in the optical flow are observed in (c).

Let $L_0(i)$ and $D_0(i)$ be the magnitude and direction, respectively, of the optical flow between the $i^{th}$ and $i + 1^{th}$ frames in the input video. Similarly, let $L_1(i)$ and $D_1(i)$ be the magnitude and direction, respectively, of the optical flow between the $i^{th}$ and $i + 1^{th}$ frames in the inpainted video. Both $L$ and $D$ are vectorized using lexicographical ordering. Then the temporal consistency between $i^{th}$ and $i + 1^{th}$ frames is given by the Pearson's correlation coefficient $r(i)$ as follows [61].

$$r(i) = \frac{1}{l-1} \sum_{j=1}^{l} \frac{(K_0^j(i) - \bar{K}_0)(K_1^j(i) - \bar{K}_1)}{\sigma_0(i)\sigma_1(i)}, \tag{8.17}$$

where $K$ can be the vector of the magnitude ($L$) or direction ($D$) of optical flow, $\bar{K}$ and $\sigma$ are mean and standard deviation of $K$, respectively, and $l$ represents the length of $K$. The value $r(i) = +1$ indicates perfect positive correlation, $r(i) = -1$ indicates perfect negative correlation while $r(i) = 0$ corresponds to no correlation between the vectors. The average value of $r$ for all the pairs of adjacent frames then gives the temporal consistency between the input and the inpainted videos. A higher average value of $r$ indicates higher temporal consistency. An example of temporal consistency in terms of optical flow is shown in Figure 8.30.

We now present an objective comparison of our method with frame-by-frame auto-inpainting in Table 8.6 in terms of the NR VQA measures viz. blockiness, blurriness and sudden local change, along with the temporal consistency measure discussed above. A video with higher blockiness and blur has higher value of the blockiness and blurriness metrics [30], respectively. For a temporally plausible video, the sudden local change [105, 106] is less while the temporal consistency measure has a higher value. From Table 8.6 we observe that the proposed method

**Table 8.6:** Comparison of proposed method with frame-by-frame auto-inpainting, in terms of blockiness (**A**), blurriness (**B**), sudden local change (**C**) and temporal consistency in optical flow's direction (**D**) and magnitude (**E**)

| Video | Proposed Method | | | | | Frame-by-Frame Auto-inpainting | | | | |
|---|---|---|---|---|---|---|---|---|---|---|
| | A | B | C | D | E | A | B | C | D | E |
| Video1 (Fig. 8.26) | 0.1125 | 5.1020 | 1.0737 | 0.9529 | 0.7501 | 0.1296 | 5.1073 | 1.3126 | 0.5064 | 0.2496 |
| Video2 (Fig. 8.27) | 0.1034 | 4.1261 | 1.5459 | 0.6671 | 0.9604 | 0.1270 | 4.2057 | 1.9463 | 0.1978 | 0.4148 |
| Video3 (Fig. 8.28) | 0.2975 | 4.3382 | 1.2908 | 0.9979 | 0.5424 | 0.2292 | 4.3666 | 1.5322 | 0.1862 | 0.6134 |
| Video4 (Fig. 8.29) | 0.1582 | 3.1264 | 2.0559 | 0.5821 | 0.9654 | 0.1662 | 3.1586 | 2.7768 | 0.2301 | 0.9381 |

performs better in terms of blockiness, sudden local change and temporal consistency, which is in accordance with the results in Figures 8.26–8.29.

The implementation details along with the timing information are presented as follows. The cracked region detection in reference frames and independent inpainting of newly detected cracked pixels are implemented in Matlab. For frames of size $270 \times 360$ (for example, the video corresponding to Figure 8.26), the inpainting of reference frames takes nearly 1.5–2 s. This includes the time required for tracking and inpainting from previously detected cracked regions (about 0.6 s) followed by initial detection, refinement and inpainting of the newly detected cracked pixels. The first frame, however, required about 4.5 s for initial detection, refinement and inpainting. Note that the size of the frame here is $270 \times 360$. Subsequent (non-reference) frames take about 0.08 s to complete tracking and inpainting from previously detected cracked regions, which is very fast when compared to independent inpainting of each frame. In this case, a considerable amount of time is required for the inpainting operation in reference frames. In real-time, this introduces a lag in the video.

Although major computational steps are implemented in C (Matlab MEX), our implementation is not an optimized version but a proof of concept of the method discussed in this section. Having said that, we are optimistic about the implementation of our method for mobile phones in order to use the method directly on videos captured onsite. This is because of the quick inpainting of subsequent (non-reference) frames. A real-time on-the-fly inpainting of the video frames could be possible with an implementation optimized for the hardware of mobile phones.

## 8.5    CONCLUSION

In this chapter, we have discussed three approaches for crack detection and also provided a method to auto-inpaint cracks in videos. Our first approach for crack detection is simple and makes use of order-statistics and density-based filters. This method is effective for detecting cracks, yet it is a primitive pixel-based approach. Our second method is a patch-based approach that performs crack detection by comparing patches using SVD. Our third crack-detection approach makes use of tolerant edit distance for comparing patches and performs more accurate crack detection in comparison to previous approaches. This approach is extended to perform automatic inpainting of video frames by making use of the scale invariant feature transform (SIFT) and homography. We also provided the temporal consistency measure to quantify the quality of the inpainted video. The reported results suggest the efficacy of our method to auto-inpaint the cracked regions in videos captured at heritage sites. In the next chapter, we discuss the challenges and future directions in digital heritage reconstruction using super-resolution and inpainting.

# CHAPTER 9

# Challenges and Future Directions

In this chapter, we discuss the challenges and future directions involved in using super-resolution and inpainting for digital heritage reconstruction. The chapter provides brief insights of these challenges, and is meant to serve as fodder that will inspire the readers to pursue them in the future.

## MOTION-BASED 3D SUPER-RESOLUTION

In this book, we have discussed techniques for image resolution enhancement and object completion using super-resolution and automatic inpainting. As mentioned in earlier chapters, these techniques provide images with enhanced resolution with no missing regions and can therefore be used as preliminary steps for generating 3D models by making use of multiple images. One can also create 3D models of heritage objects using multi-view point cloud data (PCD) acquired by laser scanners. These multiple views can be registered to obtain the complete 360-degree digitized models. However, the process aligns the available multi-view data and does not provide any enhancement in the resolution, which is a challenging task. Since the classical multi-view super-resolution techniques have been successfully applied for enhancing the resolution of 2D images, as a future direction one may explore the use of PCD captured from multiple views, not only to register but also to obtain high spatial resolution of 3D. Thus, substantially less PCD data acquired from relatively cheaper laser scanners can be used for generating high quality 3D models.

## AUTOMATIC ALIGNMENT OF CAPTURED MULTI-VIEW 3D DATA

In the above paragraph we discussed exploring the use of PCD, captured from multiple viewpoints using laser scanners, for performing motion-based 3D super-resolution, where we mentioned that the 3D digitization can be done by aligning the PCD. However, the alignment, i.e., automatic registration of the PCD itself, is challenging. This process involves two steps: (a) an initial rough registration using manually provided correspondences and (b) fine-tuning the initial registration using the iterated-closest-point algorithm. The process becomes cumbersome when hundreds of views of an object are captured for generating a high-quality 3D model. Estimating correspondences in 2D images by making use of descriptors like SIFT, etc., has been well established. This should provide motivation for developing techniques to automate the process of initial registration of 3D point cloud data which is presently done manually.

## RENDERING 3D IN HIGH-RESOLUTION WITH COLOR INFORMATION

While we may explore the use of multi-view PCD for generating high-quality 3D models, one may note that the laser scanners usually provide only the depth information. For realistic rendering of the generated 3D models, the color information in high-resolution is also required. Thus, rendering 3D in high-resolution with color information is a challenge. Although 3D estimated using multiple images has the color information, it suffers in quality in terms of resolution. For rendering purpose, data from these two sources viz. laser scanner and the color information from depth estimated using multiple images, can be merged together. However, when the resolution difference between the data obtained from these sources is large, which happens frequently, then the merging is not seamless. In applications such as creating immersive walkthrough systems, it is essential to have data obtained from the two sources to be merged/fused seamlessly so as to provide a realistic rendering to the users. This can be a motivation toward future directions in the development of algorithms for automatic merging of high-resolution laser-scanned data with low-resolution 3D data estimated using multiple color photographs.

## INPAINTING BASED ON SEMANTICS

In this book, we have developed auto-inpainting schemes to avoid human intervention. Inpainting is performed by making use of exemplars obtained by estimating patch similarity using a distance metric. However, we do not consider semantic information when performing inpainting, which is also the case with existing inpainting approaches. One may extend our work to include semantics to obtain better inpainting. Note that we human beings have a great ability to not only identify various objects even if they are partially occluded, but also "imagine the completed object" from whatever is visible to us. Thus, we humans use a two-step process to visualize the complete scene: (1) use semantics to group visual information into meaningful objects and (2) if occluded, use knowledge from our past experiences to visualize the identified object in its complete form. Hence, we may explore the image inpainting problem in a machine-learning perspective where the system needs to be trained prior to performing the desired task. For example, consider a scenario wherein an occluded wall-carving depicting the story of a coronation ceremony at a heritage site is to be completed using inpainting. Here without semantic information, it is unlikely that the missing region is appropriately inpainted in which it has people and decorations depicting the completed scene of the coronation ceremony. Intuitively, one can expect a better result if the inpainting is done with knowledge of prior information that the occluded scene represents a coronation ceremony and the exemplar search can then be restricted to scenes containing similar stories. To the best of our knowledge, the inpainting of large missing regions has not been actively researched in a machine learning framework, which is worth pursuing by using deep learning approaches. Nevertheless, works in this direction have been very recently proposed in Pathak et al. [94], Rematas et al. [102] or available as an archived manuscript in Yeh et al. [136].

## REAL-TIME INPAINTING

In Chapter 8, we observed that although the detection of cracked regions and their propagation across the video frames was fast, a considerable amount of time was spent on inpainting the detected cracked regions. In general, the challenge for exemplar-based inpainting methods lies in reducing the time complexity. These methods perform several patch comparisons due to which the speed of inpainting is slow. Although one can consider the use of the PatchMatch algorithm discussed in Barnes et al. [6] to overcome this issue, it may be noted that the best source patches may not be selected due to random search. For enabling the real-time implementation of our auto-inpainting techniques, there is a need to have a faster inpainting method. Exploring this future direction will help in performing on-the-fly inpainting on the videos captured by tourists at the heritage sites.

## BENCHMARKS FOR QUALITY ASSESSMENT OF INPAINTING

For comparing an inpainted result, a reference is usually unavailable due to which the standard criteria for quantitative comparison viz. peak signal-to-noise ratio (PSNR), root mean squared error (RMSE), structural similarity (SSIM) index, feature similarity (FSIM) index, etc., cannot be used. In such a case, the referenceless measures like the Blind/Referenceless spatial Image QUality Evaluator (BRISQUE) proposed in Mittal et al. [82], as well as the inpainted image quality assessment (IIQA) measure proposed in Trung et al. [120] are used. However, assigning subjective scores with the help of human observers is so far the most reliable way to quantify the quality of inpainted images, as well as for comparing the results of different inpainting methods. Apart from being subjective, such a comparison involves time-consuming exercises. Yet one may note that the inpainting community continues to rely on subjective comparison of the inpainted results. There is, however, a pressing need to have objective measures that can be reliably used for comparing the results of various inpainting algorithms.

# Bibliography

[1] M. V. Afonso and J. M. R. Sanches. Blind inpainting using $l_0$ and total variation regularization. *IEEE Trans. Image Processing*, 24(7), pp. 2239–2253, July 2015. DOI: 10.1109/TIP.2015.2417505. 9

[2] L. Alvarez, P.-L. Lions, and J.-M. Morel. Image selective smoothing and edge detection by nonlinear diffusion. ii. *SIAM J. Numer. Anal.*, 29, pp. 845–866, June 1992. DOI: 10.1137/0729052. 7

[3] T. Amano. Correlation based image defect detection. In *Proc. of the 18th International Conference on Pattern Recognition*, volume 01 of *ICPR '06*, pp. 163–166, Washington, DC, 2006. IEEE Computer Society. DOI: 10.1109/icpr.2006.419. xvii, 11, 14, 99, 100, 101, 107, 108, 116

[4] S. Arya and D. M. Mount. Approximate nearest neighbor queries in fixed dimensions. In *Proc. 4th Annual ACM-SIAM Symp. on Discrete Algorithms*, pp. 271–280, 1993. 25, 41

[5] S. Baker and T. Kanade. Limits on super-resolution and how to break them. *IEEE Trans. Pattern Analy. Machine Intell.*, 24(9), pp. 1167–1183, Sept. 2002. DOI: 10.1109/tpami.2002.1033210. 4

[6] C. Barnes, E. Shechtman, A. Finkelstein, and D. B. Goldman. Patchmatch: A randomized correspondence algorithm for structural image editing. *ACM Trans. Graphics*, 28(3), pp. 24:1–24:11, July 2009. DOI: 10.1145/1531326.1531330. 78, 133

[7] N. Behzad and D. Qarizadah. The man who helped blow up the bamiyan buddhas. *BBC News, Asia*, 12 March 2015. http://www.bbc.com/news/world-asia-31813681 1

[8] M. Bertalmio, G. Sapiro, V. Caselles, and C. Ballester. Image inpainting. In *Proc. 27th Annual Conf. Computer Graphics and Interactive Techniques*, pp. 417–424, New York, NY, 2000. ACM Press/Addison-Wesley Publishing Co. DOI: 10.1145/344779.344972. 7, 8, 9, 48, 67

[9] M. Brown and D. G. Lowe. Automatic panoramic image stitching using invariant features. *Int. J. Comput. Vision*, 74(1), pp. 59–73, Aug. 2007. DOI: 10.1007/s11263-006-0002-3. 118

[10] N. Cai, Z. Su, Z. Lin, H. Wang, Z. Yang, and B. W.-K. Ling. Blind inpainting using the fully convolutional neural network. *The Visual Computer*, pp. 1–13, 2015. DOI: 10.1007/s00371-015-1190-z. 9, 10

[11] E. J. Candès. The restricted isometry property and its implications for compressed sensing. *Comptes Rendus Mathematique*, 346(9–10), pp. 589–592, May 2008. 26

[12] E. J. Candès and J. Romberg. $\ell_1$-magic-recovery of sparse signals via convex programming, 2005. http://users.ece.gatech.edu/~justin/l1magic/#code 29, 63, 76

[13] E. J. Candès and M. B. Wakin. An introduction to compressive sampling. *IEEE Signal Processing Magazine*, 25(2), pp. 21–30, March 2008. DOI: 10.1109/msp.2007.914731. 62, 76

[14] T. Chan and L. Vese. Active contours without edges. *IEEE Trans. Image Processing*, 10(2), pp. 266–277, Feb. 2001. DOI: 10.1109/83.902291. 115

[15] R.-C. Chang, Y.-L. Sie, S.-M. Chou, and T. K. Shih. Photo defect detection for image inpainting. In *Proc. of the 7th IEEE International Symposium on Multimedia*, ISM '05, pp. 403–407, Washington, DC, 2005. IEEE Computer Society. DOI: 10.1109/ism.2005.91. 11

[16] S. Chaudhuri. *Super-Resolution Imaging*. Kluwer Academic Publishers, Norwell, MA, 2001. DOI: 10.1007/b117840. 3

[17] S. Chaudhuri and M. V. Joshi. *Motion-Free Super-Resolution*. Springer, 2005. ISBN 9780387258904. DOI: 10.1007/b136446. 3

[18] D. Chen and R. J. Plemmons. *The Birth of Numerical Analysis*, chapter Nonnegativity Constraints in Numerical Analysis, pp. 109–138. World Scientific, 2009. 26, 47, 49, 51

[19] B. Cornelis, T. Ružić, E. Gezels, A. Dooms, A. Pižurica, L. Platiša, J. Cornelis, M. Martens, M. D. Mey, and I. Daubechies. Crack detection and inpainting for virtual restoration of paintings: The case of the ghent altarpiece. *Signal Processing*, 93(3), pp. 605–619, 2013. Image Processing for Digital Art Work. DOI: 10.1016/j.sigpro.2012.07.022. 12, 14

[20] A. Criminisi, P. Pérez, and K. Toyama. Object removal by exemplar-based inpainting. In *Proc. IEEE Int. Conf. on Computer Vision and Pattern Recognition*, volume 2, p. 721, Los Alamitos, CA, 2003. IEEE Computer Society. DOI: 10.1109/cvpr.2003.1211538. 8, 9, 48

[21] A. Criminisi, P. Pérez, and K. Toyama. Region filling and object removal by exemplar-based image inpainting. *IEEE Trans. Image Processing*, 13, pp. 1200–1212, 2004. DOI:

10.1109/tip.2004.833105. 7, 8, 9, 12, 47, 48, 49, 50, 54, 55, 57, 59, 72, 74, 75, 93, 101, 105, 106, 107, 115, 118, 122

[22] Z. Cui, H. Chang, S. Shan, B. Zhong, and X. Chen. Deep network cascade for image super-resolution. In *Computer Vision—ECCV 2014: 13th European Conference, Part V*, pp. 49–64, 2014. DOI: 10.1007/978-3-319-10602-1_4. 6

[23] S. Darabi, E. Shechtman, C. Barnes, D. B. Goldman, and P. Sen. Image melding: Combining inconsistent images using patch-based synthesis. *ACM Trans. Graphics*, 31(4), pp. 82:1–82:10, July 2012. DOI: 10.1145/2185520.2335433. 78

[24] S. Deerwester, S. T. Dumais, G. W. Furnas, T. K. Landauer, and R. Harshman. Indexing by latent semantic analysis. *J. American. Soc. for Information Sci.*, 41(6), pp. 391–407, 1990. DOI: 10.1002/(sici)1097-4571(199009)41:6%3C391::aid-asi1%3E3.0.co;2-9. 102, 103

[25] B. Dong, H. Ji, J. Li, Z. Shen, and Y. Xu. Wavelet frame based blind image inpainting. *Applied and Computational Harmonic Analysis*, 32(2), pp. 268–279, 2012. DOI: 10.1016/j.acha.2011.06.001. 9

[26] C. Dong, C. C. Loy, K. He, and X. Tang. Learning a deep convolutional network for image super-resolution. In *Computer Vision—ECCV 2014: 13th European Conference, Part IV*, pp. 184–199, 2014. DOI: 10.1007/978-3-319-10593-2_13. 6

[27] C. Dong, C. C. Loy, K. He, and X. Tang. Image super-resolution using deep convolutional networks. *IEEE Trans. Pattern Analy. Machine Intell.*, 38(2), pp. 295–307, Feb. 2016. DOI: 10.1109/tpami.2015.2439281. 6

[28] D. L. Donoho. Compressed sensing. *IEEE Trans. Information Theory*, 52(4), pp. 1289–1306, Apr. 2006. DOI: 10.1109/tit.2006.871582. 26

[29] C. Emerson, N. Lam, and D. Quattrochi. Multi-scale fractal analysis of image texture and patterns. *Photogrammetric Eng. and Remote Sensing*, 65(1), pp. 51–62, Jan. 1999. 85

[30] M. Farias and S. Mitra. No-reference video quality metric based on artifact measurements. In *Proc. Int. Conf. Image Processing*, volume 3, pp. III–141–4, 2005. DOI: 10.1109/icip.2005.1530348. 126, 127

[31] S. Farsiu, D. Robinson, M. Elad, and P. Milanfar. Fast and robust multi-frame super-resolution. *IEEE Trans. Image Processing*, 13(10), pp. 1327–1344, Oct. 2004. DOI: 10.1109/tip.2004.834669. 4

[32] R. Fattal. Image upsampling via imposed edge statistics. *ACM Trans. Graphics*, 26(3), p. 95, July 2007. DOI: 10.1145/1239451.1239546. 6, 33

[33] O. Faugeras and F. Lustman. Motion and structure from motion in a piece-wise planar environment. Technical Report RR-0856, INRIA, June 1988. DOI: 10.1142/s0218001488000285. 119

[34] M. A. Fischler and R. C. Bolles. Random sample consensus: a paradigm for model fitting with applications to image analysis and automated cartography. *Commun. ACM*, 24(6), pp. 381–395, June 1981. DOI: 10.1145/358669.358692. 118

[35] G. Freedman and R. Fattal. Image and video upscaling from local self-examples. *ACM Trans. Graphics*, 28(3), pp. 1–10, Apr. 2011. DOI: 10.1145/1944846.1944852. 6

[36] W. T. Freeman, T. R. Jones, and E. C. Pasztor. Example-based super-resolution. *IEEE Computer Graphics and Applications*, 22(2), pp. 56–65, Mar. 2002. DOI: 10.1109/38.988747. 5

[37] Y. Furukawa and J. Ponce. Accurate, dense, and robust multiview stereopsis. *IEEE Trans. Pattern Analy. Machine Intell.*, 32(8), pp. 1362–1376, Aug. 2010. DOI: 10.1109/tpami.2009.161. 1

[38] P. P. Gajjar and M. V. Joshi. A fast approach to the learning based super-resolution using autoregressive model prior and wavelet prior. In *Proc. of the 14th National Conference on Communications*, pp. 229–233, 2008. 51

[39] P. P. Gajjar and M. V. Joshi. New learning based super-resolution: Use of DWT and IGMRF prior. *IEEE Trans. Image Processing*, 19(5), pp. 1201–1213, May 2010. DOI: 10.1109/tip.2010.2041408. 6

[40] M. Ghorai, P. Purkait, and B. Chanda. A fast video inpainting technique. In *Pattern Recognition and Machine Intelligence*, volume 8251 of *Lecture Notes in Computer Science*, pp. 430–436. Springer Berlin Heidelberg, 2013. DOI: 10.1007/978-3-642-45062-4_59. 12

[41] M. Ghorai, S. Mandal, and B. Chanda. Image completion assisted by transformation domain patch approximation. In *Proc. of the 2014 Indian Conference on Computer Vision Graphics and Image Processing*, ICVGIP '14, pp. 66:1–66:8, New York, NY, 2014. ACM. DOI: 10.1145/2683483.2683549. 9, 14

[42] D. Glasner, S. Bagon, and M. Irani. Super-resolution from a single image. In *Proc. 12th IEEE Conf. Computer Vision*, pp. 349–356, Mar. 2009. DOI: 10.1109/iccv.2009.5459271. 6, 13, 19, 20, 21, 23, 24, 25, 29, 30, 31, 33, 72, 74, 77, 78, 79, 81, 82

[43] R. C. Gonzalez and R. E. Woods. *Digital Image Processing*. Addison-Wesley Longman Publishing Co., Inc., Boston, MA, 2nd ed., 2001. ISBN 0201180758. 94

[44] Google Images, [Online] Available: `http://www.images.google.com`, Mar. 2012. 88, 105

[45] H. Grossauer. A combined pde and texture synthesis approach to inpainting. In *8th European Conference on Computer Vision (ECCV 2004)*, volume 4, pp. 214–224, 2004. DOI: 10.1007/978-3-540-24671-8_17. 8

[46] A. Grün, F. Remondino, and L. Zhang. Photogrammetric reconstruction of the great buddha of bamiyan, afghanistan. *The Photogrammetric Record*, 19(107), pp. 177–199, sept. 2004. DOI: 10.1111/j.0031-868x.2004.00278.x. 1

[47] R. C. Hardie, K. J. Barnard, and E. E. Armstrong. Joint map registration and high-resolution image estimation using a sequence of undersampled images. *IEEE Trans. Image Processing*, 6(12), pp. 1621–1633, Dec. 1997. DOI: 10.1109/83.650116. 28

[48] P. F. Harrison. Gimp resynthesizer plugin, 2011. `http://www.logarithmic.net/pfh/resynthesizer`. 78

[49] R. Hartley and A. Zisserman. *Multiple View Geometry in Computer Vision*. Cambridge University Press, New York, NY, 2nd ed., 2003. DOI: 10.1017/cbo9780511811685. 118

[50] K. He and J. Sun. Statistics of patch offsets for image completion. In *12th European Conference on Computer Vision (ECCV 2012)*, pp. 16–29, 2012. DOI: 10.1007/978-3-642-33709-3_2. 78

[51] J.-B. Huang, S. B. Kang, N. Ahuja, and J. Kopf. Image completion using planar structure guidance. *ACM Trans. Graphics*, 33(4), pp. 129:1–129:10, July 2014. DOI: 10.1145/2601097.2601205. 9, 72, 78, 81

[52] J.-B. Huang, S. B. Kang, N. Ahuja, and J. Kopf. Dataset for image completion using planar structure guidance, 2014. `https://sites.google.com/site/jbhuang0604/publications/struct_completion` DOI: 10.1145/2601097.2601205. 78

[53] J.-B. Huang, A. Singh, and N. Ahuja. Single image super-resolution from transformed self-exemplars. In *Proc. IEEE Int. Conf. on Computer Vision and Pattern Recognition*, 2015. DOI: 10.1109/cvpr.2015.7299156. 6

[54] T. S. Huang and R. Y. Tsai. Multi-frame image restoration and registration. *Advances in Computer Vision and Image Processing*, 1, pp. 317–339, 1984. 4

[55] M. Irani and S. Peleg. Improving resolution by image registration. *CVGIP: Graphical Model and Image Processing*, 53(3), pp. 231–239, Apr. 1991. DOI: 10.1016/1049-9652(91)90045-1. 4

[56] S. Jeannin and A. Divakaran. MPEG-7 visual motion descriptors. *IEEE Trans. Circuits and Syst. for Video Tech.*, 11(6), pp. 720–724, 2001. DOI: 10.1109/76.927428. 117

[57] D. J. Jobson, Z. Rahman, and G. A. Woodell. Properties and performance of a center/surround retinex. *IEEE Trans. Image Processing*, 6(3), pp. 451–462, Mar. 1997. DOI: 10.1109/83.557356. 84

[58] M. V. Joshi, L. Bruzzone, and S. Chaudhuri. A model-based approach to multiresolution fusion in remotely sensed images. *Geoscience and Remote Sensing, IEEE Transactions on*, 44(9), pp. 2549–2562, Sept. 2006. DOI: 10.1109/tgrs.2006.873340. 49

[59] S. Katahara and M. Aoki. Face parts extraction windows based on bilateral symmetry of gradient direction. In *Computer Analysis of Images and Patterns*, volume 1689, pp. 834–834. Springer Berlin Heidelberg, 1999. DOI: 10.1007/3-540-48375-6_59. 84

[60] M. Kazhdan, M. Bolitho, and H. Hoppe. Poisson surface reconstruction. In *Proc. of the 4th Eurographics Symposium on Geometry Processing*, SGP '06, pp. 61–70, Aire-la-Ville, Switzerland, 2006. Eurographics Association. 2, 47, 80

[61] J. F. Kenney. *Mathematicals of Statistics*. Van Nostrand, 1954. 127

[62] N. Khatri and M. V. Joshi. Image super-resolution: Use of self-learning and gabor prior. In *11th Asian Conference on Computer Vision (ACCV 2012)*, pp. 413–424, 2012. DOI: 10.1007/978-3-642-37431-9_32. 17, 20, 25, 27, 28, 33, 75

[63] N. Khatri and M. V. Joshi. Efficient self-learning for single image upsampling. In *22nd International Conference in Central Europe on Computer Graphics, Visualization and Computer Vision (WSCG 2014)*, pp. 1–8, 2014. 17, 37, 38, 39, 40

[64] N. Komodakis and G. Tziritas. Image completion using efficient belief propagation via priority scheduling and dynamic pruning. *IEEE Trans. Image Processing*, 16(11), pp. 2649–2661, Nov. 2007. DOI: 10.1109/tip.2007.906269. 78

[65] P. D. Kovesi. MATLAB and Octave functions for computer vision and image processing. Centre for Exploration Targeting, School of Earth and Environment, The University of Western Australia, 2005. Available from: http://www.csse.uwa.edu.au/~pk/research/matlabfns /robust/ransacfithomography.m 118

[66] C. Kwan, B. Ayhan, G. Chen, J. Wang, B. Ji, and C.-I. Chang. A novel approach for spectral unmixing, classification, and concentration estimation of chemical and biological agents. *Geoscience and Remote Sensing, IEEE Transactions on*, 44(2), pp. 409–419, Feb. 2006. DOI: 10.1109/tgrs.2005.860985. 49, 51

[67] Y. Lee and A. Fam.   An edge gradient enhancing adaptive order statistic filter. *IEEE Trans. Acoustics, Speech and Signal Proc.*, 35(5), pp. 680–695, May 1987. DOI: 10.1109/tassp.1987.1165188. 94

[68] Y.-H. Lee and S. Tantaratana. An adaptive edge enhancing order statistic filter. In *Proc. IEEE Int. Conf. on Acoustics, Speech and Signal Processing*, volume 3, pp. 1518–1521, Apr. 1988. DOI: 10.1109/icassp.1988.196892. 94

[69] P. Legrand. Local regularity and multifractal methods for image and signal analysis. In *Scaling, Fractals and Wavelets*. Wiley, Jan. 2009. DOI: 10.1002/9780470611562.ch11. 85

[70] Z. Lin and H.-Y. Shum. Fundamental limits of reconstruction-based superresolution algorithms under local translation. *IEEE Trans. Pattern Analy. Machine Intell.*, 26(1), pp. 83–97, Jan. 2004. DOI: 10.1109/tpami.2004.1261081. 4

[71] D. Liu, Z. Wang, B. Wen, J. Yang, W. Han, and T. S. Huang.   Robust single image super-resolution via deep networks with sparse prior. *IEEE Trans. Image Processing*, 25(7), pp. 3194–3207, July 2016. DOI: 10.1109/tip.2016.2564643. 6

[72] D. G. Lowe.   Distinctive image features from scale-invariant keypoints.   *International Journal of Computer Vision*, 60(2), pp. 91–110, Nov. 2004. DOI: 10.1023/b:visi.0000029664.99615.94. 113, 118

[73] X. Lu, H. Yuan, Y. Yuan, P. Yan, L. Li, and X. Li. Local learning-based image superresolution. In *Multimedia Signal Processing (MMSP), 2014 IEEE 16th International Workshop on*, pp. 1–5, Oct. 2011. DOI: 10.1109/mmsp.2011.6093843. 20

[74] B. D. Lucas and T. Kanade. An iterative image registration technique with an application to stereo vision. In *Proc. 7th Int. Joint Conf. on AI*, pp. 674–679, 1981. 127

[75] H. Q. Luong, T. Ružić, A. Pižurica, and W. Philips. Single-image super-resolution using sparsity constraints and non-local similarities at multiple resolution scales. In *Proc. SPIE*, volume 7723, pp. 772305–772305–8, 2010. DOI: 10.1117/12.854437. 13

[76] Q. Luong, T. Ruzic, A. Pizurica, and W. Philips.   Single-image super-resolution using sparsity constraints and non-local similarities at multiple resolution scales. In *Proc. of the Society Of Photo-Optical Instrumentation Engineers (SPIE)*, volume 7723, p. 8, 2010. DOI: 10.1117/12.854437. 6

[77] Y. Ma, S. Soatto, J. Kosecka, and S. S. Sastry. *An Invitation to 3-D Vision: From Images to Geometric Models*. Springer Verlag, 2003. ISBN 0387008934. DOI: 10.1007/978-0-387-21779-6. 119, 120

[78] L. W. MacDonald. *Digital Heritage: Applying Digital Imaging to Cultural Heritage*. Elsevier, 2006. ISBN 9780750661836. DOI: 10.1093/llc/fqn002. 13

[79] D. Martin, C. Fowlkes, D. Tal, and J. Malik. A database of human segmented natural images and its application to evaluating segmentation algorithms and measuring ecological statistics. In *Proc. 8th IEEE Conf. Computer Vision*, volume 2, pp. 416–423, July 2001. DOI: 10.1109/iccv.2001.937655. 37

[80] S. Masnou and J.-M. Morel. Level lines based disocclusion. In *Proc. Int. Conf. Image Processing*, volume 3, pp. 259–263, Oct. 1998. DOI: 10.1109/icip.1998.999016. 7, 8, 9, 48, 67

[81] S. McCloskey, M. Langer, and K. Siddiqi. Removal of partial occlusion from single images. *IEEE Trans. Pattern Analy. Machine Intell.*, 33(3), pp. 647–654, march 2011. DOI: 10.1109/tpami.2010.187. 9

[82] A. Mittal, A. K. Moorthy, and A. C. Bovik. No-reference image quality assessment in the spatial domain. *IEEE Trans. Image Processing*, 21(12), pp. 4695–4708, Dec. 2012. DOI: 10.1109/tip.2012.2214050. 31, 133

[83] U. M. Munshi and B. B. Chaudhuri. *Multimedia Information Extraction and Digital Heritage Preservation*. Platinum Jubilee series. World Scientific, 2011. ISBN 9789814307253. DOI: 10.1142/9789814307260. 13

[84] K. Nasrollahi and T. B. Moeslund. Super-resolution: A comprehensive survey. *Machine Vision and Applications*, 25(6), pp. 1423–1468, Aug. 2014. DOI: 10.1007/s00138-014-0623-4. 3

[85] A. Newson, A. Almansa, M. Fradet, Y. Gousseau, and P. Pérez. Video inpainting of complex scenes. *SIAM Journal on Imaging Sciences, Society for Industrial and Applied Mathematics*, 7(4), pp. 1993–2019, 2014. DOI: 10.1137/140954933. 12

[86] M. M. Oliveira, B. Bowen, R. Mckenna, and Y. sung Chang. Fast digital image inpainting. In *Proc. of the International Conference on Visualization, Imaging and Image Processing (VIIP 2001)*, pp. 261–266. ACTA Press, 2001. 8

[87] M. G. Padalkar and M. V. Joshi. Auto-inpainting heritage scenes: a complete framework for detecting and infilling cracks in images and videos with quantitative assessment. *Machine Vision and Applications*, 26(2), pp. 317–337, 2015. DOI: 10.1007/s00138-015-0661-6. 17

[88] M. G. Padalkar, M. V. Joshi, M. A. Zaveri, and C. M. Parmar. Exemplar based inpainting using autoregressive parameter estimation. In *Proc. of the International Conference on Signal, Image and Video Processing*, ICSIVP'12, pp. 154–160, IIT Patna, India, Jan. 2012. 17, 93, 97

[89] M. G. Padalkar, M. V. Vora, M. V. Joshi, M. A. Zaveri, and M. S. Raval. Identifying vandalized regions in facial images of statues for inpainting. In *New Trends in Image Analysis and Processing—ICIAP 2013*, volume 8158 of *Lecture Notes in Computer Science*, pp. 208–217. Springer Berlin Heidelberg, Sept. 2013. DOI: 10.1007/978-3-642-41190-8_23. 17

[90] M. G. Padalkar, M. A. Zaveri, and M. V. Joshi. SVD based automatic detection of target regions for image inpainting. In J.-I. Park and J. Kim, Eds., *Computer Vision—ACCV 2012 Workshops*, volume 7729 of *Lecture Notes in Computer Science*, pp. 61–71. Springer Berlin Heidelberg, 2013. DOI: 10.1007/978-3-642-37484-5. 17, 116

[91] M. G. Padalkar, M. V. Joshi, and N. Khatri. Simultaneous inpainting and super-resolution using self-learning. In X. Xie, M. W. Jones, and G. K. L. Tam, Eds., *Proc. of the British Machine Vision Conference (BMVC)*, pp. 105.1–105.12. BMVA Press, Sept. 2015. DOI: 10.5244/c.29. 17

[92] S. C. Park, M. K. Park, and M. G. Kang. Super-resolution image reconstruction: A technical overview. *Signal Processing Magazine, IEEE*, 20(3), pp. 21–36, May 2003. DOI: 10.1109/msp.2003.1203207. 3, 4

[93] C. M. Parmar, M. V. Joshi, M. S. Raval, and M. A. Zaveri. Automatic image inpainting for the facial images of monuments. In *Proc. of Electrical Engineering Centenary Conference 2011*, pp. 415–420, 14–17 Dec. 2011. 11

[94] D. Pathak, P. Krähenbühl, J. Donahue, T. Darrell, and A. Efros. Context encoders: Feature learning by inpainting. In *Proc. IEEE Int. Conf. on Computer Vision and Pattern Recognition*, 2016. (accepted). 133

[95] K. A. Patwardhan, G. Sapiro, and M. Bertalmío. Video inpainting under constrained camera motion. *IEEE Trans. Image Processing*, 16(2), pp. 545–553, 2007. DOI: 10.1109/tip.2006.888343. 12, 15, 118, 126

[96] P. Pérez, M. Gangnet, and A. Blake. Poisson image editing. In *Proc. 30th Annual Conf. Computer Graphics and Interactive Techniques*, pp. 313–318, New York, NY, 2003. ACM. DOI: 10.1145/1201775.882269. 8

[97] P. Pérez, M. Gangnet, and A. Blake. Poisson image editing. *ACM Trans. Graphics*, 22(3), pp. 313–318, July 2003. DOI: 10.1145/882262.882269. 8, 9, 10, 47, 48, 51, 52, 83, 88

[98] P. Perona and J. Malik. Scale-space and edge detection using anisotropic diffusion. *IEEE Trans. Pattern Analy. Machine Intell.*, 12, pp. 629–639, July 1990. DOI: 10.1109/34.56205. 7, 84

[99] N. Petkov. Biologically motivated computationally intensive approaches to image pattern recognition. *Future Generation Computer Systems*, 11(4–5), pp. 451–465, Aug. 1995. DOI: 10.1016/0167-739x(95)00015-k. 27

[100] P. Purkait and B. Chanda. Digital restoration of damaged mural images. In *Proc. of the 8th Indian Conference on Computer Vision, Graphics and Image Processing*, ICVGIP '12, pp. 49:1–49:8, New York, NY, 2012. ACM. DOI: 10.1145/2425333.2425382. 9, 14

[101] P. Purkait and B. Chanda. Super resolution image reconstruction through bregman iteration using morphologic regularization. *IEEE Trans. Image Processing*, 21(9), pp. 4029–4039, sept. 2012. DOI: 10.1109/tip.2012.2201492. 71

[102] K. Rematas, C. Nguyen, M. Fritz, and T. Tuytelaars. Novel views of objects from a single image. *IEEE Trans. Pattern Analy. Machine Intell.*, 2016. (to appear). DOI: 10.1109/tpami.2016.2601093. 133

[103] C. Rother, V. Kolmogorov, and A. Blake. "rabcut:" Interactive foreground extraction using iterated graph cuts. *ACM Trans. Graphics*, 23(3), pp. 309–314, Aug. 2004. DOI: 10.1145/1015706.1015720. 115

[104] D. L. Ruderman and W. Bialek. Statistics of natural images: Scaling in the woods. *Physical Review Letters*, 73(6), pp. 814–817, Aug. 1994. DOI: 10.1103/physrevlett.73.814. 6

[105] M. Saad, A. Bovik, and C. Charrier. Blind prediction of natural video quality. *IEEE Trans. Image Processing*, 23(3), pp. 1352–1365, Mar. 2014. DOI: 10.1109/tip.2014.2299154. 126, 127

[106] M. A. Saad and A. C. Bovik. Blind quality assessment of videos using a model of natural scene statistics and motion coherency. In *Asilomar Conference on Signals, Systems, and Computers*, pp. 332–336, Nov. 2012. DOI: 10.1109/acssc.2012.6489018. 126, 127

[107] G. Sapiro. *Geometric Partial Differential Equations and Image Analysis*. Cambridge University Press, 2001. ISBN 9780521790758. DOI: 10.1017/cbo9780511626319. 65

[108] S. Scholes. Mcconkie ranch petroglyphs near vernal, Utah, Nov. 2011. https://www.youtube.com/watch?v=jmewuqEXTK8 [Accessed 01 Sept. 2014]. 124

[109] Q. Shan, Z. Li, J. Jia, and C.-K. Tang. Fast image/video upsampling. In *ACM SIGGRAPH Asia 2008 Papers*, SIGGRAPH Asia '08, pp. 153:1–153:7, New York, NY, 2008. ACM. DOI: 10.1145/1457515.1409106. 41, 42, 43

[110] T. Shibata, A. Iketani, and S. Senda. Image inpainting based on probabilistic structure estimation. In *10th Asian Conference on Computer Vision (ACCV 2010)*, volume III, pp. 109–120, Berlin, Heidelberg, 2011. Springer-Verlag. DOI: 10.1007/978-3-642-19318-7_9. 9

[111] T. Shih, N. Tang, and J.-N. Hwang. Exemplar-based video inpainting without ghost shadow artifacts by maintaining temporal continuity. *IEEE Trans. Circuits and Syst. for Video Tech.*, 19(3), pp. 347–360, Mar. 2009. DOI: 10.1109/tcsvt.2009.2013519. 12

[112] T. Sikora.  MPEG digital video-coding standards. *Signal Processing Magazine, IEEE*, 14(5), pp. 82–100, 1997. DOI: 10.1109/79.618010. 117

[113] F. Stanco, S. Battiato, and G. Gallo.  *Digital Imaging for Cultural Heritage Preservation: Analysis, Restoration, and Reconstruction of Ancient Artworks*.  Digital Imaging and Computer Vision. CRC Press, 2011. ISBN 9781439821749. 14

[114] J. Sun, Z. Xu, and H.-Y. Shum.  Image super-resolution using gradient profile prior.  In *Proc. IEEE Int. Conf. on Computer Vision and Pattern Recognition*, pp. 1–8, 2008. DOI: 10.1109/cvpr.2008.4587659. 6

[115] T. Tamaki, H. Suzuki, and M. Yamamoto.  String-like occluding region extraction for background restoration. *International Conference on Pattern Recognition*, 3, pp. 615–618, 2006. ISSN 1051-4651. DOI: 10.1109/icpr.2006.1082. 11

[116] C. V. L. The University of Tokyo.  Virtual Asukakyo, 2013. http://www.cvl.iis.u-tokyo.ac.jp/research/virtual-asukakyo/ 1

[117] J. Tian and K.-K. Ma.  Stochastic super-resolution image reconstruction. *Journal of Visual Communication and Image Representation*, 21(3), pp. 232–244, Apr. 2010. DOI: 10.1016/j.jvcir.2010.01.001. 28

[118] R. Tibshirani, G. Walther, and T. Hastie.  Estimating the number of clusters in a data set via the gap statistic. *Journal of Royal Stat. Soc., B*, 63(2), pp. 411–423, 2001. DOI: 10.1111/1467-9868.00293. 86

[119] R. Timofte, , V. De Smet, and L. Van Gool. A+: Adjusted anchored neighborhood regression for fast super-resolution. In *12th Asian Conference on Computer Vision (ACCV 2014)*, Nov. 2014. DOI: 10.1007/978-3-319-16817-3_8. 6

[120] D. T. Trung, A. Beghdadi, and M.-C. Larabi.  Perceptual quality assessment for color image inpainting.  In *Proc. Int. Conf. Image Processing*, pp. 398–402, Sept. 2013. DOI: 10.1109/icip.2013.6738082. 133

[121] N. Turakhia, R. Shah, and M. Joshi.  Automatic crack detection in heritage site images for image inpainting.  In *8th Indian Conference on Computer Vision, Graphics and Image Processing (ICVGIP)*, pp. 68, 2012. DOI: 10.1145/2425333.2425401. 11, 116

[122] A. Turiel, G. Mato, N. Parga, and J. pierre Nadal. The self-similarity properties of natural images resemble those of turbulent flows. *Physical Review Letters*, 80, pp. 1098–1101, Feb. 1998. DOI: 10.1103/physrevlett.80.1098. 6

[123] P. Vandewalle, S. Süsstrunk, and M. Vetterli.  A Frequency Domain Approach to Registration of Aliased Images with Application to Super-Resolution. *EURASIP Journal on*

*Applied Signal Processing (special issue on Super-resolution)*, 2006: 71459 (14 pages), 2006. DOI: 10.1155/asp/2006/71459. 5

[124] M. Varma and A. Zisserman. Classifying images of materials: Achieving viewpoint and illumination independence. In *7th European Conference on Computer Vision (ECCV 2002)*, pp. 255–271, 2002. DOI: 10.1007/3-540-47977-5_17. 83, 85, 86

[125] V. Štruc and N. Pavešić. *Photometric normalization techniques for illumination invariance*, pp. 279–300. IGI-Global, 2011. DOI: 10.4018/978-1-61520-991-0.ch015. 84

[126] R. A. Wagner and M. J. Fischer. The string-to-string correction problem. *J. ACM*, 21(1), pp. 168–173, Jan. 1974. DOI: 10.1145/321796.321811. 111

[127] M. E. Wall, A. Rechtsteiner, and L. M. Rocha. *Singular Value Decomposition and Principal Component Analysis*, chapter Singular value decomposition and principal component analysis, pp. 91–109. Kluwer: Norwell, 2003. DOI: 10.1007/0-306-47815-3_5. 103

[128] Z. Wang, A. C. Bovik, H. R. Sheikh, and E. P. Simoncelli. Image quality assessment: From error visibility to structural similarity. *IEEE Trans. Image Processing*, 13(4), pp. 600–612, Apr. 2004. ISSN 1057-7149. http://dx.doi.org/10.1109/TIP.2003.819861 DOI: 10.1109/tip.2003.819861. 30

[129] Z. Wang, F. Zhou, and F. Qi. Inpainting thick image regions using isophote propagation. In *Proc. Int. Conf. Image Processing*, pp. 689–692, Oct. 2006. DOI: 10.1109/icip.2006.312428. 8

[130] Z. Wang, D. Liu, J. Yang, W. Han, and T. Huang. Deep networks for image super-resolution with sparse prior. In *Proc. 2015 IEEE Conf. Computer Vision*, pp. 370–378, Dec. 2015. DOI: 10.1109/iccv.2015.50. 6

[131] Y. Wexler, E. Shechtman, and M. Irani. Space-time completion of video. *IEEE Trans. Pattern Analy. Machine Intell.*, 29(3), pp. 463–476, March 2007. DOI: 10.1109/tpami.2007.60. 9, 47, 72, 75, 78, 124, 126

[132] J. Wu and Q. Ruan. Object removal by cross isophotes exemplar-based inpainting. In *Proc. of the 18th International Conference on Pattern Recognition (ICPR'06)*, volume 3, pp. 810–813, Washington, DC, 2006. IEEE Computer Society. DOI: 10.1109/icpr.2006.884. 8

[133] J. Xie, L. Xu, and E. Chen. Image denoising and inpainting with deep neural networks. In P. Bartlett, F. Pereira, C. Burges, L. Bottou, and K. Weinberger, Eds., *Advances in Neural Information Proc. Systems 25, Proc. 26nd Annual Conf. on Neural Information Proc. Systems*, pp. 350–358, 2012. 9

[134] M. Yan. Restoration of images corrupted by impulse noise and mixed gaussian impulse noise using blind inpainting. *SIAM J. on Imaging Sciences*, 6(3), pp. 1227–1245, 2013. DOI: 10.1137/12087178x. 9

[135] J. Yang, J. Wright, T. S. Huang, and Y. Ma. Image super-resolution via sparse representation. *IEEE Trans. Image Processing*, 19(11), pp. 2861–2873, Nov. 2010. DOI: 10.1109/tip.2010.2050625. 6, 13

[136] R. Yeh, C. Chen, T. Y. Lim, M. Hasegawa-Johnson, and M. N. Do. Semantic image inpainting with perceptual and contextual losses. *arXiv preprint arXiv:1607.07539*, 2016. 133

[137] Q. Yuan, L. Zhang, and H. Shen. Regional spatially adaptive total variation super-resolution with spatial information filtering and clustering. *IEEE Trans. Image Processing*, 22(6), pp. 2327–2342, June 2013. DOI: 10.1109/tip.2013.2251648. 71

[138] L. Zhang, L. Zhang, X. Mou, and D. Zhang. Fsim: A feature similarity index for image quality assessment. *IEEE Trans. Image Processing*, 20(8), pp. 2378–2386, Aug. 2011. DOI: 10.1109/tip.2011.2109730. 30

[139] M. Zontak and M. Irani. Internal statistics of a single natural image. In *Proc. IEEE Int. Conf. on Computer Vision and Pattern Recognition*, pp. 977–984, June 2011. DOI: 10.1109/cvpr.2011.5995401. 20

[140] Q. Zou, Y. Cao, Q. Li, Q. Mao, and S. Wang. Cracktree: Automatic crack detection from pavement images. *Pattern Recognition Letters*, 33(3), pp. 227–238, 2012. DOI: 10.1016/j.patrec.2011.11.004. 11, 14, 105

# Authors' Biographies

## MILIND G. PADALKAR

**Milind G. Padalkar** received a B.E. degree in information technology from the University of Mumbai, India, in 2008 and an M.Tech. degree in computer engineering from Sardar Vallabhbhai National Institute of Technology, Surat, India, in 2010. Currently he is pursuing a Ph.D. degree in information and communication technology and worked as a Junior Research Fellow from 2011 to 2016 (in the Indian Digital Heritage project) at Dhirubhai Ambani Institute of Information and Communication Technology, Gandhinagar, India. His research interests include image processing and computer vision.

## MANJUNATH V. JOSHI

**Manjunath V. Joshi** received a B.E. degree from the University of Mysore, Mysore, India, and M.Tech. and Ph.D. degrees from the Indian Institute of Technology Bombay (IIT Bombay), Mumbai, India. Currently, he is serving as a Professor with the Dhirubhai Ambani Institute of Information and Communication Technology, Gandhinagar, India. He has been involved in active research in the areas of signal processing, image processing and computer vision. He has co-authored a book entitled *Motion-Free Super Resolution* (Springer, New York). Dr. Joshi was a recipient of the Outstanding Researcher Award in Engineering Section by the Research Scholars Forum of IIT Bombay. He was also a recipient of the Best Ph.D. Thesis Award by Infineon India and the Dr. Vikram Sarabhai Award for the year 2006–2007 in the field of information technology constituted by the Government of Gujarat, India.

## NILAY L. KHATRI

**Nilay L. Khatri** received his B.Tech. degree in electronics and communication engineering from the Institute of Technology, Nirma University, Ahmedabad, India, in 2008 and an M.Tech. degree in information and communication technology from Dhirubhai Ambani Institute of Information and Communication Technology (DA-IICT), Gandhinagar, India, in 2011. He has worked as a Junior Research Fellow in the Indian Digital Heritage project at DA-IICT, Gandhinagar, India, from 2011 to 2014. His research interests include image processing and 3D computer vision. He is particularly interested in 3D content generation from 2D data by incorporating human interactions in computer vision algorithms. Currently, he is working as an Image Processing Research Engineer at Jekson-Vision, India.

Printed in the United States
by Baker & Taylor Publisher Services